GPS and Google Earth for Development

How to create, share and collaborate with maps on the net

First Edition.
September 2012.

Santiago Arnalich
Julio Urruela

arnalich

water and habitat

GPS and Google Earth for Development

How to create, share and collaborate with maps on the net

First Edition.
September 2012.

ISBN: 978-84-616-0235-3

© Arnalich. Water and habitat

If you wish to use part of the contents of this book, get in touch with us at
publicaciones@arnalich.com

Errata at: www.arnalich.com/dwnl/xgoopsen.doc

arnalich

water and habitat

To Aldara and her grandmother.

Our gratitude to all those who, directly or indirectly, contributed to the making of this book, thanks go to:

- Loreto Romeral for her initial explorations on the subject for her end of course dissertation (ASCHI).
- Adam Schneider for the creation and his dedication to GPSvisualizer.
- Amelia Jiménez, Mary Brown and Maxim Fortin for their patience with the proofreading.
- Federica Bonvin for volunteering for the photographs.
- Carlos Velasco of www.ingenieria-velasco.es , who is always willing to give a hand.

Index

I. Introduction and tools

About this book 7
A no-frills no-fuss GIS 8
How this book is organized 9
What you need 10

1. Google Earth   11

Presenting Google Earth...12
Downloading and installing Google Earth 13
Handling the Earth Globe..16
An example of an online map 18
Exercise 1: Opening a Google Earth file...19
Layers or maps? Peeling the onion 21
Exercise 2: Activating and deactivating layers....................................22
Exercise 3: Saving and compressing layers 24
Exercise 4: Adding points, lines and areas...26

2. GPS and coordinates   31

What is a GPS?.... 32
The workings of the system 35
Types of data you can collect 36
Projections and coordinates...41
Navigating with coordinates 45
Sources of error and precision 46
Cold starting a GPS 49

An indispensable ritual: configuration of the GPS 50
Working with the GPS 54
Uploading and downloading data 60
Navigating with Google Earth in real time 68
Using Google Earth offline 69
Using third party maps 71
Exercise 5. Overlaying maps 74

3. Create   77

Exercise 6. Downloading and converting GPS data 78
Exercise 7. Adding color to the position markers...87
Exercise 8. Adding a legend 91
Exercise 9. Adding color according to quantities 95
Exercise 10. Classifying in folders 97
Exercise 11. Adding sub-folders 100
Exercise 12. Creating maps for a website 102
Exercise 13. Displaying data within the info balloons 104

4. Share   107

Before sharing 108
A mailing list or a group..109
A blog 110
A website 111
File hosting 112

5. Collaborate   113

Before collaborating 114
Collaborating with Google Drive...115

About the authors **119**
Bibliography.. **121**

About this book

This book gives you the necessary information so that you can document and communicate what happens on the ground in an effective manner after just a weekend, or perhaps two, of reading.

Many things will change, new ways of working will emerge and new programs will become available. We don't intend this book to be the latest cutting edge manual nor one that covers specific programs. Rather we ask **you to concentrate on the process of creating, sharing and collaborating** with the information regarding maps and how they can be useful for your cause.

It's intended to be:

> **99 % fat free.** Without extensive explanations or never-ending demonstrations. Only what's necessary has been included.

> **Simple**. One of the frequent causes of failure is that detail and over-strictness end up intimidating the user and so things don't get done. Risking insult, the explanations assume almost nothing to be obvious.

> **Chronological**. It approximates the logical order in which you would do the project.

> **Practical**. With plenty of examples.

> **Self-contained**. With this book, a GPS and a computer conected to the internet, you will have everything you need.

Basically, the content of this book is aimed at building a **younger brother of a Geographic Information System**, with the same basic functions but without the necessity of specific programs, huge knowledge or laborious learning.

By keeping things simple you will have at your disposal an easy, quick and effective way to comunicate and colaborate inside and outside your organization with people like you, who have an interest in what happens but don't have access to the necessary tools.

A no-frills, no-fuss GIS

One of the principal limitations of Geographic Information Systems (GIS) is that they end up being limited to a small "elite" of people that manage them and channel everyone else's requests be within the organization or from other organizations.

These elite is often saturated with requests, with limited patience and generally you have to physically visit them in faraway places to obtain information, as sharing GIS information on a website is not an easy task. In addition, sometimes the people and organizations develop a kind of territorial syndrome about the data they have.

In any case, it's very likely that they are not gathering all the data that you need, which means that, even though its information is useful in a general sense, it doesn't solve the problem of mapping your specific data.

Using a collaborative and easy to learn system like Google Earth can help promote:

- The **participation** of aid workers, institutions as well as the beneficiaries in the collection and analysis of the information without intermediaries.

- The **transparency** with a tool in which you can see the state of things and inform the public in a participative manner.

- The creation of awareness in the **public opinion**. One good example is the layer about the genocide done by the United States Darfur Holocaust Museum that you can download here: www.ushmm.org/maps/crisisindarfur.kmz

How this book is organized

1. **It's progressive**. The exercises follow the logical order of the process.

2. It has **warnings**:

ATTENTION! This symbol warns of the most frequent errors to prevent it "blowing up in your face".

BEWARE, TIME WASTING! As in all IT programs and also in real life, it is very easy to get bogged down in work that has no value so that even "the snails will be ahead of you".

3. It has **routes** that look like this: "> Tools/ Options". The bar indicates that there has been a jump in the menu, so that route is equivalent to clicking on "Tools" in the general menu and "Options" in the one that drops down:

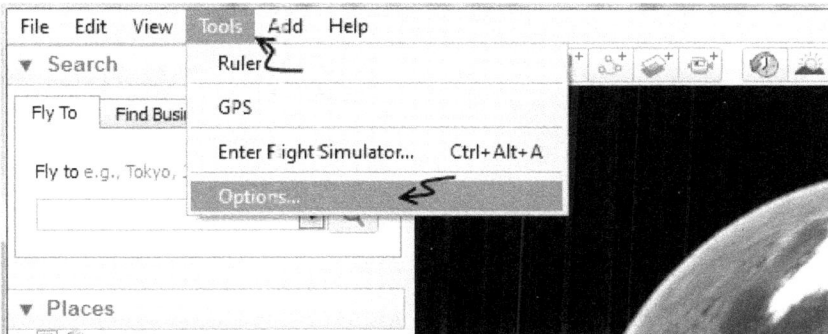

4. It has **symbols** to make reading easier:

Download necessary! In order to carry out the work in hand you will need to download the file indicated. You can download all the content at once here: www.arnalich.com/dwnl/goopsen/allcontent.zip

Brief **explanations** and useful observations.

What you need

To do the things proposed here you will need:

- Access to a **computer** and a basic knowledge of how to use it.

- Access to **internet**, at least from time to time so that you can convert the files with GPSvisualizer and publish your maps. Even though there are ways of working without connection with programs like KMLCSV or CSV2KML, it's the distribution through the internet and the collaboration that makes the maps worthwhile.

- A **GPS** and a **connection cable** to download the data to the computer.

- Some software: **Google Earth** and **GPSBabel**

Brand Disclaimer

Brand names are included in this book. However we do not endorse any commercial products, processes, or services and you may find similar products to those mentioned that suit you better. These are mentioned as examples only to illustrate a process. Also, if a particular brand name is not mentioned, this does not mean or imply that the product is unsatisfactory.

1

Google Earth

Presenting Google Earth

In this book we use Google Earth. It seems like a good choice because it's free, easy to learn and use, it's available in many languages and has a host of people already familiarized with it.

The quality of the satellite images is very good and they constantly get better. Also, since the hurricane Katrina, Google Earth has a tradition of renewing and improving the images of a zone after a disaster in a matter of days.

Although it has a Pro version (free for the NGOs after certain paperwork), the free version and the Pro use the same satellite images and share most of the useful functions. For everything described in this manual, the free version is adequate.

This is what the program looks like:

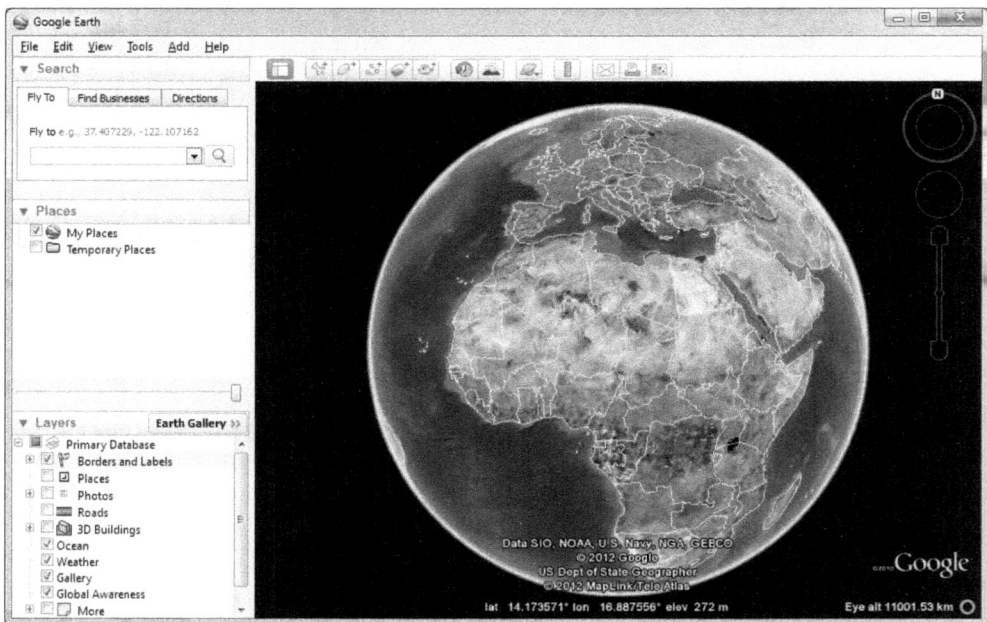

In the next section you'll be able to download it and to navigate a real map soon after.

Downloading and installing Google Earth

1. Download the program at:

   http://earth.google.com

2. Once there, look for the download button:

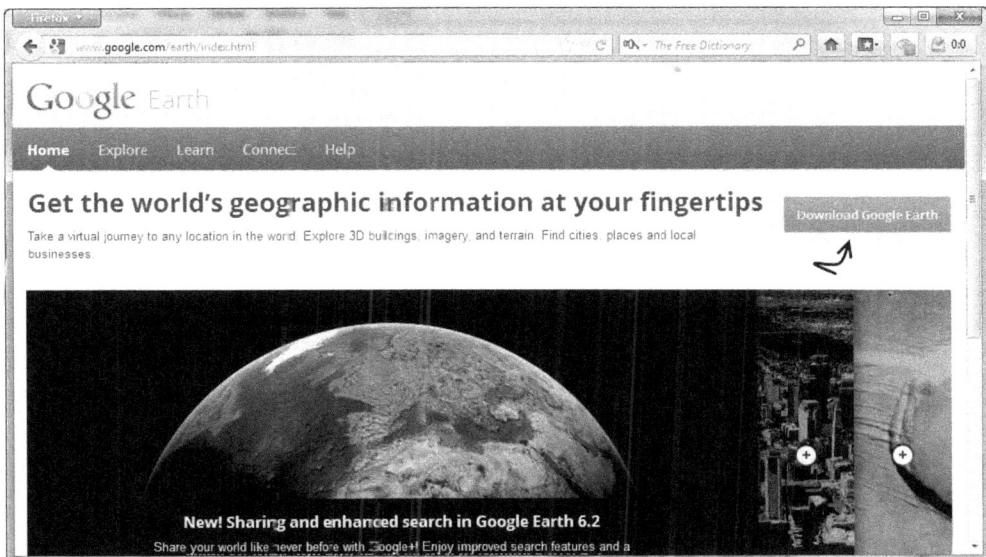

3. To download, it's not necessary to download Google Chrome or any other program. If you don't want them, remove the marks in the boxes on the page when they offer them. To start the download click Accept and Agree and download:

   Download the latest version of Google Earth for PC, Mac, or

   ☐ Include Google Chrome, a fast, free browser made for the modern web. Learn more
   ☐ Make Google Chrome my default browser.

   By installing, you agree to Google Earth's Privacy Policy .

   Google Maps/Earth Additional Terms of Service

   Last Modified: March 1, 2012

   By downloading, installing, or using the Google Earth software, accessing or using the Google Maps se (together, the "**Products**" or "**Services**"), or accessing or using any of the content available within the F

   As a condition of downloading, accessing, or using the Products, you also agree to the terms of the Go

   ▸ Customize your installation of Google Earth with advanced setup

   Agree and Download

4. If your web browser gives you the option of executing the download do so, if
not you'll have to look for the file GoogleEarthWin in your downloads folder
and double click to start it (the file may have other names such as
googleupdatesetup.exe):

5. Click install in the box:

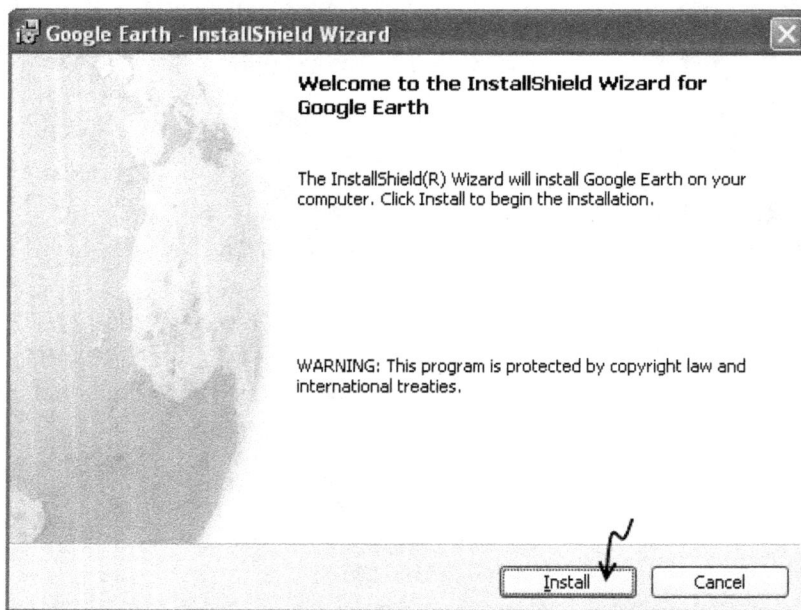

6. After a series of progress bars the installation finished box appears. If you
check the box Launch Google Earth, the program will open after pressing
Finish:

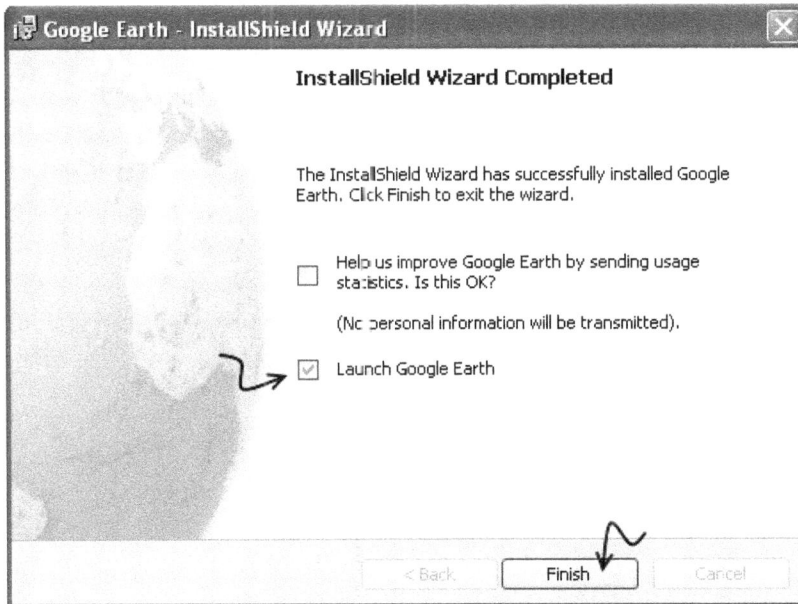

To open it afterwards in Windows, press the start button (white arrow), look between the programs for the Google Earth folder and click on Google Earth. Alternatively, you can type Google Earth in the Search programs and files box:

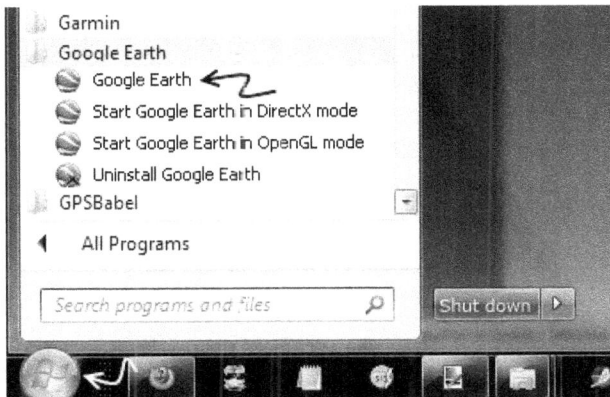

7. Leave the program open for the next section where you'll see the basic navigation in Google Earth.

Handling the Earth Globe

When you open Google Earth the Navigation Controls appear in the top right corner, which allow you to go from one place to another, zoom in or out and even see things in 3D.

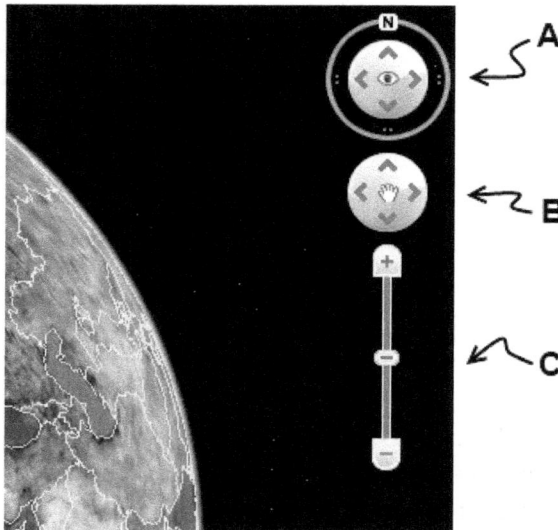

The **Direction control and 3D** (A), allows you to orientate the globe. To see what we're referring to, click on the N and, without letting go, drag it around the control until completing a circle:

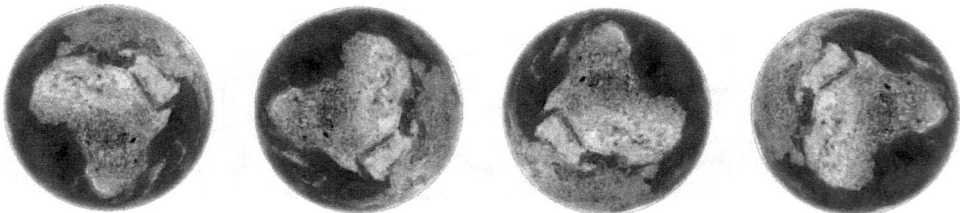

Clicking on the N, the north automatically returns to the top of the screen.

The vertical arrows of this control allow you to see the relief in 3D. The top arrow, for example, has the same effect as lifting your gaze from vertically into the ground progressively to the horizon. Looking at the ground from above you see things flat.

Lifting your gaze you start seeing at an angle and the relief appears, for example, the Canary Islands in relief against the horizon:

The horizontal arrows make the image turn as if you were turning around.

The **control pan** arrows (B) move the map from side to side and up and down without changing the vertical angle or spin.

The **Zoom Control** (C) does exactly what you would imagine. You can click on the + and – signs or drag the bar.

But the quickest way to find what you want is to tap it so that it spins like a real globe. To do this, "grab" the globe by clicking with the mouse without letting go, and drag it to your liking. Although to start with, it will move uncontrollably and it can be a bit distracting, soon you'll move it as you want with high speed.

An example of an online map

Points/Properties Maps

This is one of the most common and useful types of map. The point marks the position while the size, color, type of icon and so on depends on some property: i.e. if it's green the water of a well is dirty, if it's red its pump is broken or if it's white it's working perfectly.

This is the color code of the following map of Haiti[1] that you're going to use very soon in the first exercise:

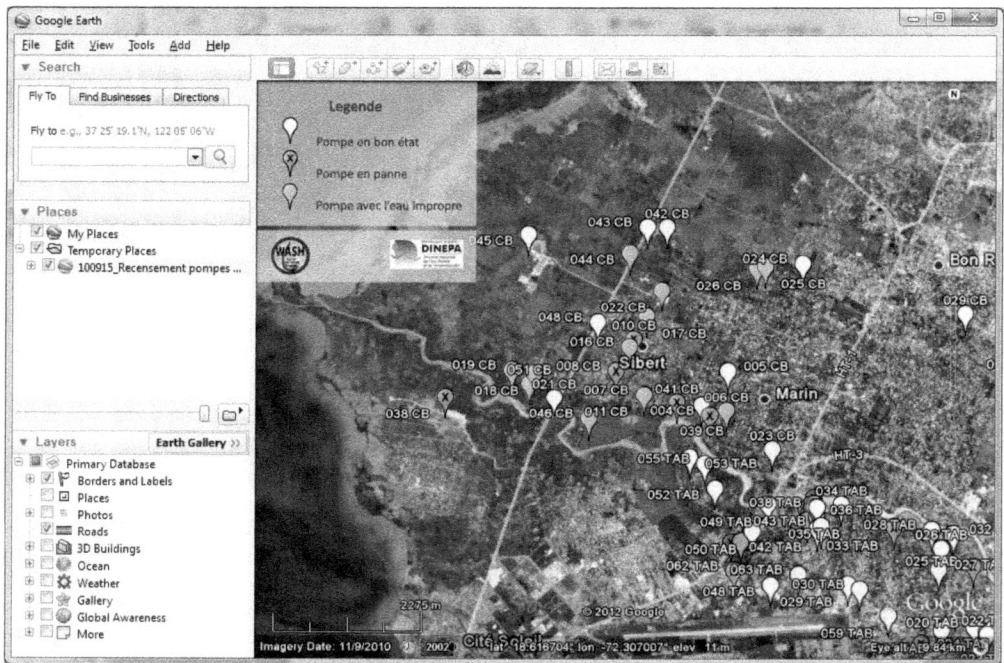

At first sight, we have an idea of the general state and where the problems are concentrated which would be much more difficult studying a list. Since we have less and less time these days and more divided attention, **communicating via maps is fundamental for important messages NOT go unnoticed.**

As you will have already imagined there are also **property-line** maps (i.e. state of roads: useable, closed…), and **property-polygon** (i.e. uses of the ground: agricultural, industrial…).

---

[1]KMZ file produced by the WASH Cluster Haiti 2010.

Opening a Google Earth file

Open the KMZ file with the condition of the wells in Haiti.

1. Download the file here:

       www.arnalich.com/dwnl/goopsen/wells.kmz

2. To open it you can click on the file and Google Earth will open with the file loaded:

    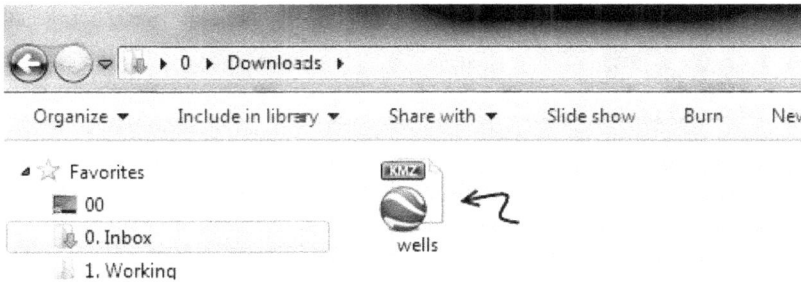

    Alternatively, you can open it from the program by following File / Open:

    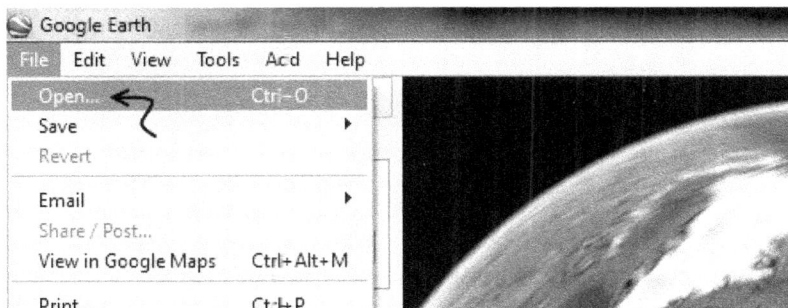

    Notice that the file appears loaded in the Places box. This box creates an index of all the information that you load:

Notice also that even though the file is called Wells, it can have a different name when it loads in Places.

3. Take the opportunity to get used to the controls if you didn't do so in the previous section (Handling the Earth Globe).

Layers or Maps? Peeling the onion

Imagine that, besides the state of the wells, you want to see where the provisional internally displaced person (IDP) camps are to prioritize the repair of the nearest water points. You can load a second file[1] to see them together:

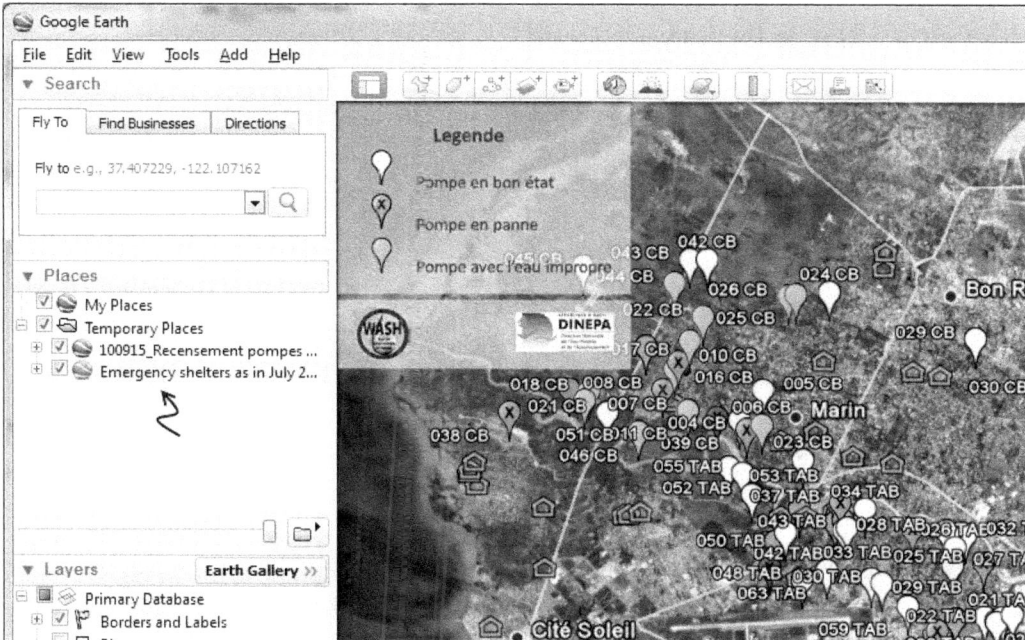

The new file is placed on top of the earlier one like a transparent layer over which the points representing the IDP camps have been drawn. You can carry on adding information with transparent **layers** until you think the **map** has what you need.

---

[1] *KMZ file produced by WASH Cluster Haiti 2010.*

Activating and deactivating layers

2

Create a map that shows the data of the IPD camps, hide the information of hospitals and show only the Wells in the area of the Croix des Bouquets without the legend.

1. Open Google Earth if you closed it.

2. Load the layer of Wells as you did in exercise 1,> File/ Open.

3. Download the 2 remaining layers:

    www.arnalich.com/dwnl/goopsen/IDP.kmz
    www.arnalich.com/dwnl/goopsen/hospitals.kml

4. Load the two layers you just downloaded.

If you hold down the **ctrl key**, you can select various layers at the same time, instead of opening them one by one. If the list of layers is very long, you can also click on the first one, press the **shift key** ↑, and then click on the last one. All the layers that you've selected will appear shaded:

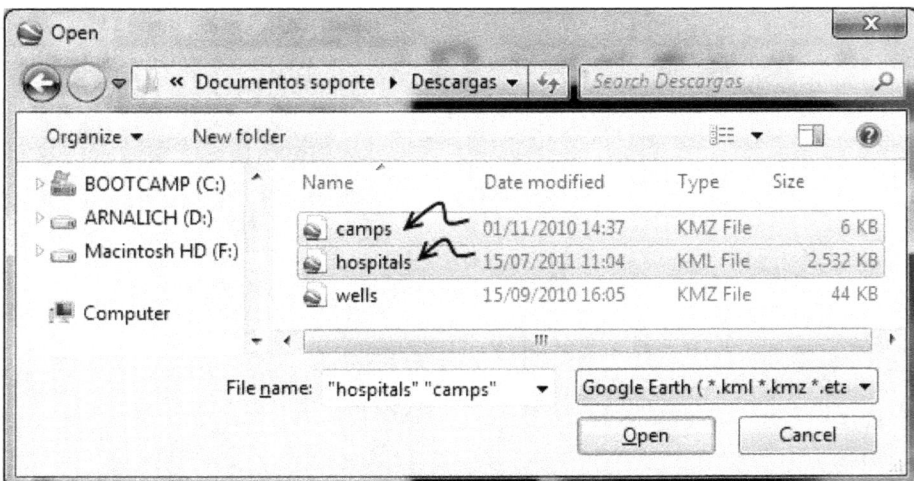

After loading the layers, the index box will look like this one:

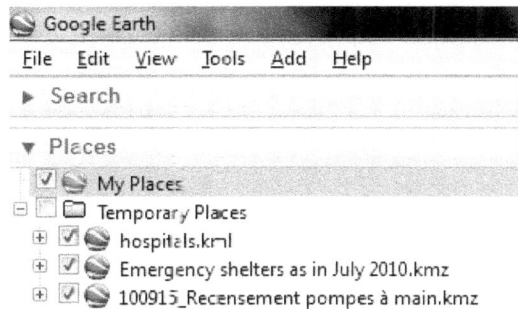

5. Make the hospital.kml layer nvisible clicking on the box to uncheck it:

6. To display only parts of the layer of Wells you have to open it to see what it contains. To do so, press the + sign at its side:

7. Once opened, you can deactivate content so it will only show the Wells in the Croix de Bouquets area without legend. The boxes should look like these:

Note that the layers that show only part of their information are in blue.

Don't close the program just yet, you will use these changes in the following exercise.

Saving and compressing layers

3

Save the KML of hospitals as KMZ.

First a brief explanation:

There are two types of files that you can use on Google Earth:

- **KML** are normal files, written in a very similar language to the html used in web pages so that it is easy to modify.

- **KMZ,** which is the abbreviation for KML-Zipped, it's the same file but compressed. That way all the personalized information that could go with the KML file (personalized icons, superimposed images, etc.) is included in just one package.

1. Carrying on from the last exercise, right click on the hospital layer and choose Save as:

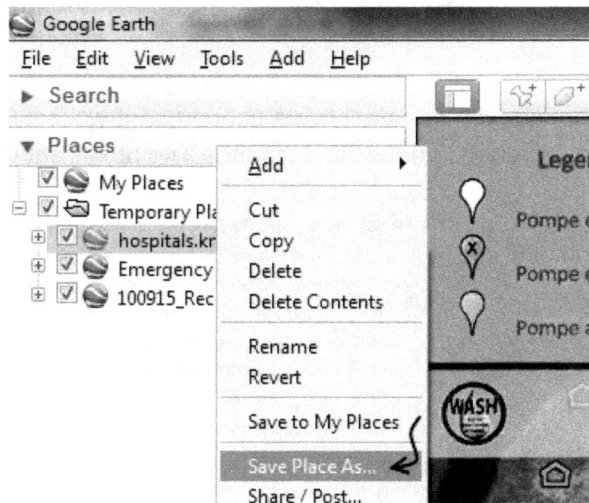

2. In the menu that opens, write hospitals2 as the name and in the Save as type drop menu select Kmz:

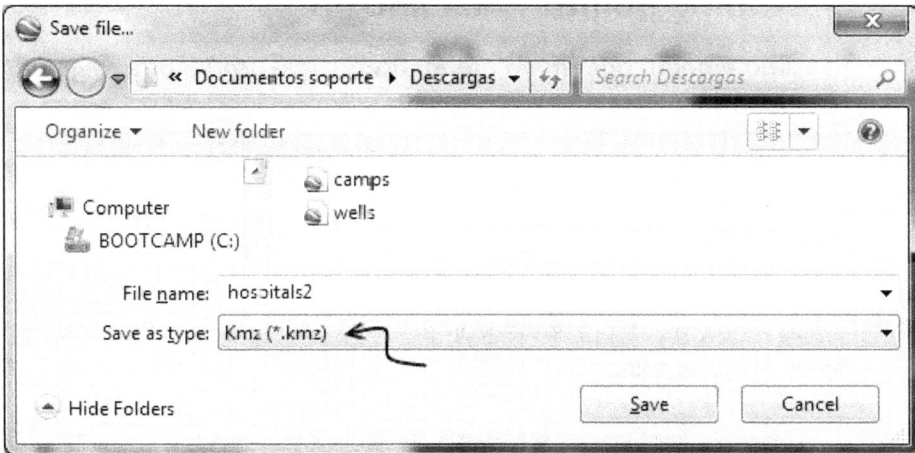

3. Go to the folder in your computer where you have saved it and notice the difference in size between the new file and the other one in Kml:

Name	Type	Size
camps	KMZ File	6 KB
hospitals	KML File	2.532 KB
hospitals2	KMZ File	128 KB
wells	KMZ File	44 KB

The file that you have saved is 20 times smaller. When the internet connection is very slow size is important. Also remember that if you want to collaborate with other people, everyone else doesn't necessarily have the same connection speed as you do.

Adding points, lines and areas

4

Add a point, determine the topographical profile of a road and point out a crop plot.

1. Open Google Earth and navigate to whatever place you feel like. The place isn't important for this exercise.

2. To add a new point, press the icon that looks like a pin and a blinking pin will appear in the screen:

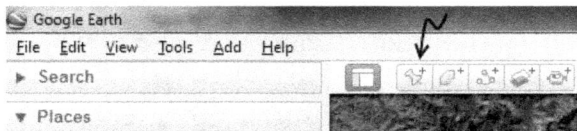

This pin can be either dragged from one place to another until you accept it or be assigned coordinates to define where it should be placed.

You can choose other symbols for the pin in the dialogue that appears by clicking the pin button next to the field Name. You can also change the color, add descriptions, etc., by accessing the options on the tabs.

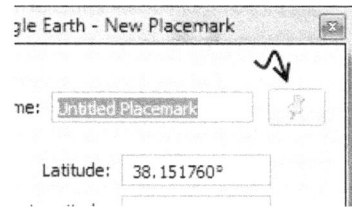

3. Now press the route icon:

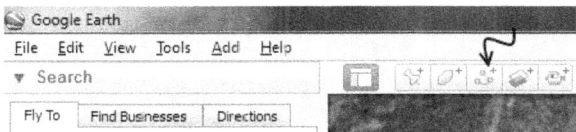

4. Click consecutive points to define any random line and press accept when you want to finish:

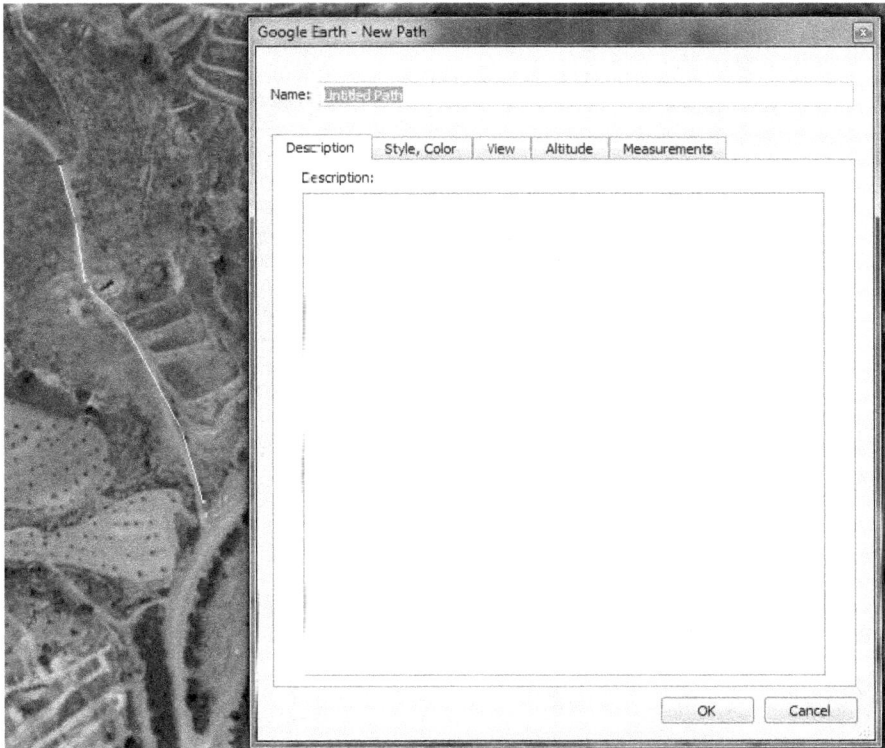

Notice that both the route and the point have been added to the Place table.

5. Right click on the route and select Show Elevation Profile:

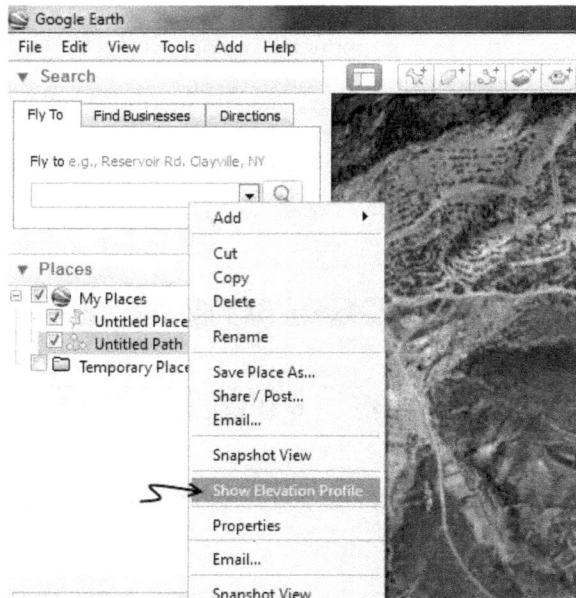

The result will be similar to this one. You can move the cursor to obtain the information of the specific red arrow point:

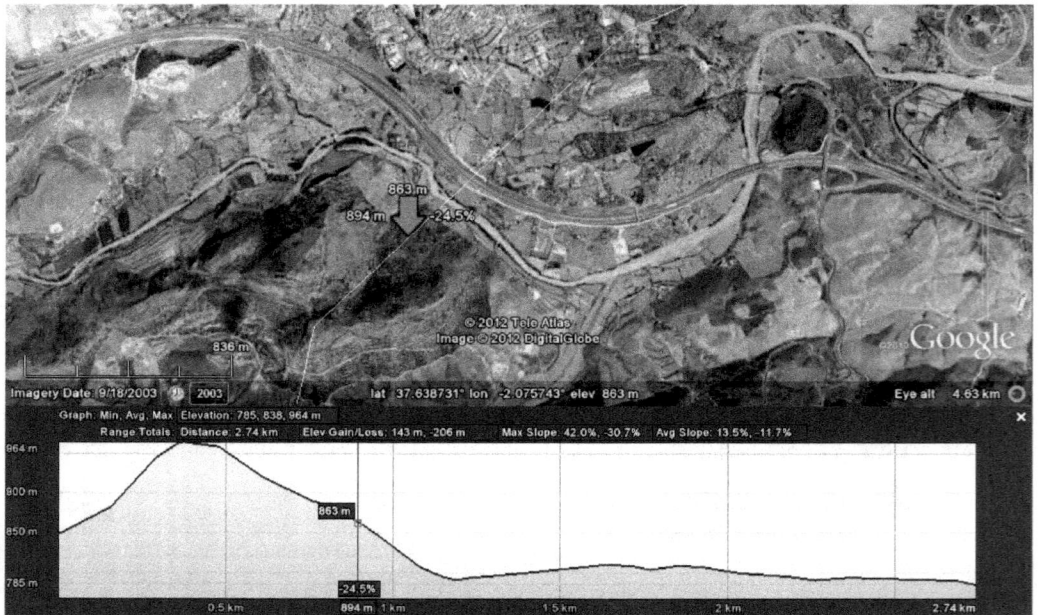

Remember that the precision of these profiles isn't enough for a lot of engineering jobs, for example, to plan a drinking water scheme.

6. Click on the area icon:

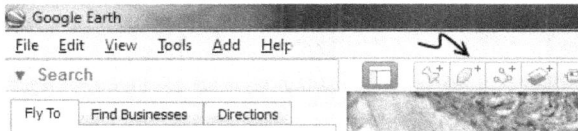

This function will add a triangle to the area with every click. It may be a little confusing at first, but just click on the points that will form the border of the area clockwise or anticlockwise, but always in the same direction and you'll get the knack of it quickly:

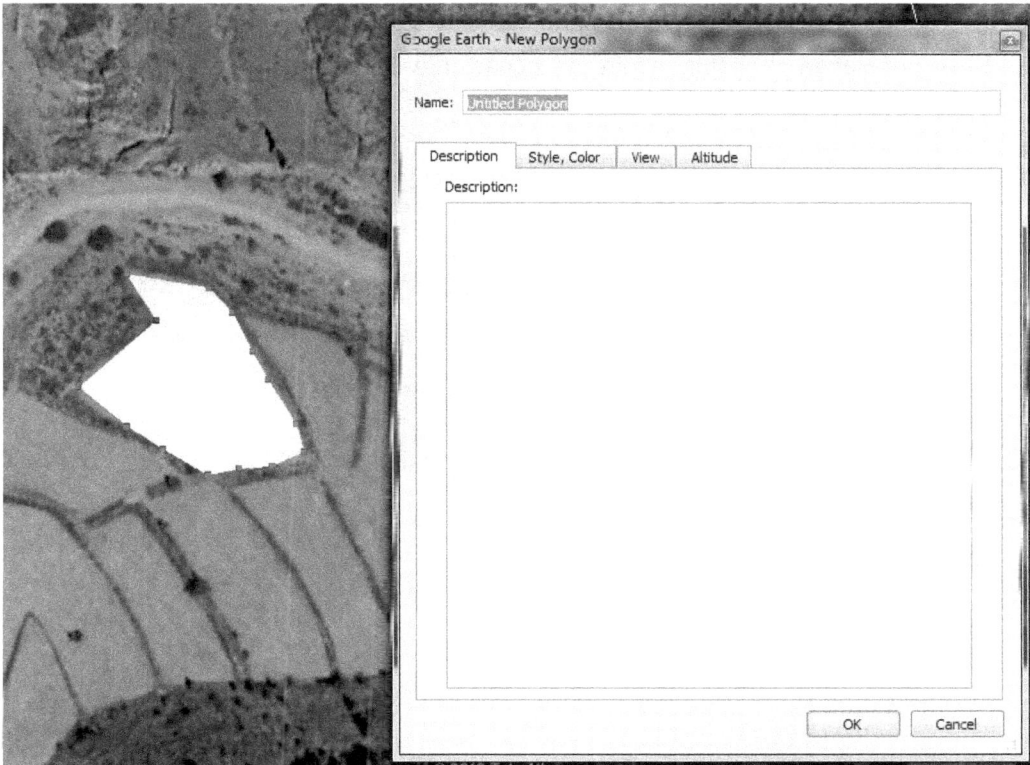

7. You can add transparency and color in the Style, Color tab:

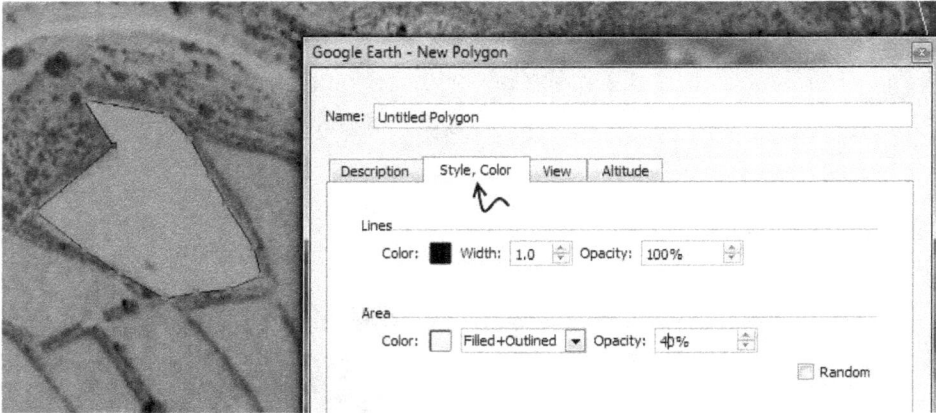

Take advantage and play around with the rest of the buttons to look for options that might be useful to you. For example, with the clock you can get images from different years. Comparing them you can see the evolution over time of the deforestation, count houses to figure out the growth rate of a population or simply avoid the occasional cloud that stops you from seeing your work area:

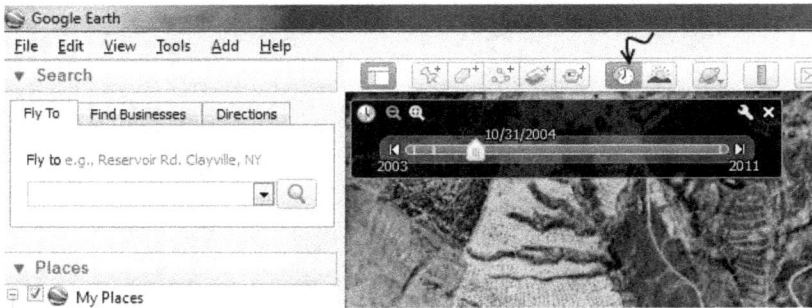

2

GPS and Coordinates

What is a GPS?

A GPS is a device that permits you to know the coordinates of a point with great precision (around 3 m for conventional coordinates). Its use is actually easier than that of a simple mobile phone. Here we are going to concentrate on you knowing what you're doing and let the manual explain its use.

Unless you use a GPS you will be limited to the information given to you by third parties or trying to recognize places in the satellite images of Google Earth. This is impractical and liable to error in cities and virtually impossible in open areas. In the context of Development Aid it is highly improbable that all the data you need will be available and up to date. In an emergency it would be impossible. You will need to be able to determine the coordinates you need by yourself!

The photo shows a traditional handheld GPS. The simplest models like this Garmin etrex H costs around $100 and have all the functions that you will need.

Although you might be tempted to go for a very complicated model, using a simple one gives you a black and white screen that you can see very well in bright sun light, simple menus and low use of batteries. A simple GPS unit is also less likely to be stolen.

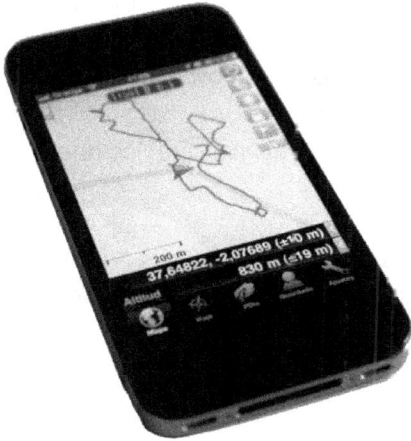

A lot of smart phones already have integrated GPS, which could be very useful in places with good coverage, reasonable call costs and speed to see satellite images. Furthermore you can take photos of places, make notes, etc. on the same machine. However…

…with these make sure that the technology doesn't end up taking all your attention, with a phone full of applications and intanglements which prevents you from seeing the reality that you want to chart.

Another option for some uses are the GPS sport watches. In the image you see one connected to one of the innumerable sports website:

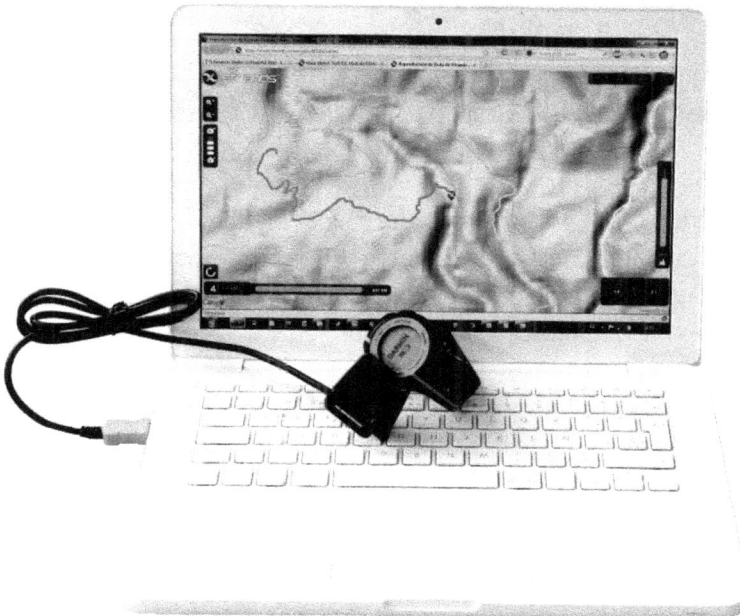

These tend to register tracks better, avoiding tracks to be shred up into new bits every time you momentarily lose signal. Also, it allows you to calculate rhythms, average speed and laps without great hassle. Another interesting application is that they calculate the approximate calorie loss, which could be useful to make more visible some of the difficulties that the population has, avoiding the coldness and inexpressivity of ways of stating problems.

For example, the Mamao spring in Haiti is less than a kilometer away from the population (which is within the European Union standards) and to begin with it would have been passed as good:

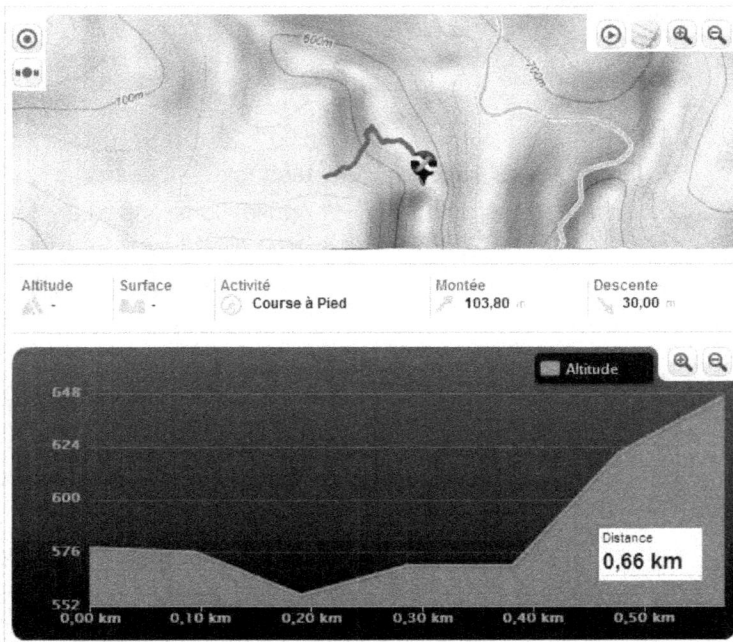

Adding that the climb to get water is at 103m helps you to get an idea, but saying that the journey to fetch water uses approximately 25% of the calorie intake of a child is much more revealing to the problem, and it helps bring down pigeonhole attitudes such as "my organization has prioritized food safety projects".

Going back to the GPS watch, one of its main problems is that they don't mark the points in a straightforward manner. Another inconvenience is that they don't incorporate maps.

The workings of the system

The position given by a GPS is calculated by triangulation with respect to a network of satellites. Each satellite emits a repetitive signal that allows the GPS to know the distance between them. Our position can be any point of the sphere made by points that comply to that distance from the satellite. Using the signal of another 2 satellites and the surface of the Earth, a unique point is obtained where all the spheres intersect. This point is our position. The more satellites the GPS can reach, the more defined this intesection point and thus the higher the precision we get.

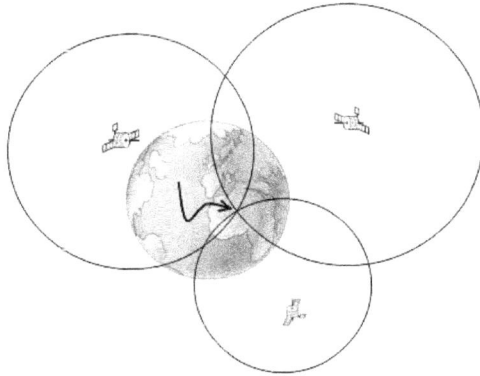

The system was created by the Defense Department of the USA initially only for military use. After a commercial plane crash with 269 passengers lost in the restricted air space in the Soviet Union, it was opened to civil use.

There is also a Russian system (GLONASS) open to civil use since 2007, a European system (Galileo) and a Chinese one (Compass) in development.

Types of data you can collect

Basically, everything that is linked to a coordinate: the position of objects, the trajectory of the roads, the borders of a plot of land, etc. An important thing to remember is that:

With a traditional GPS, **don't take the altitude data seriously** even though it diligently displays a number (for example, 879m like in the image below). According to the geometry of the satellites the errors can be of hundreds of meters.

If the GPS has a barometer, the precision is around twenty meters, which still isn't very useful other than to get an idea because on the internet you can get the levels with that degree of precision for free from a digital elevation model.

Points or waypoints: where is it?

When the GPS saves a point, what it is doing is saving it's coordinates to be able to situate it on any map. The goal is to answer the question **where is the...** (maternity, well, school...)?

You mark a way point by simply standing on the place and pressing the button. The GPS then shows the coordinates and gives a series of options, for example, to change the icon that's going to represent it. To avoid compatibility problems, it is better that the names have 6 or less characters and don't use accents or non-universal signs (ñ,Д,ф,œ...).

Naming points is generally a waste of time, and the limitation of 6 characters makes you scratch your head afterwards trying to figure out what MTNY03[1] was. If time has passed since you worked on it or it's someone else that receives the file there are going to be some wild guesses!

---

[1] *Maternity 3*

It's a lot more practical to make a note of the number of the point in a notebook along with the rest of the descriptions about the place, as you'll soon see.

If naming points is generally a waste of time, coding some type of information using icons such as *the houses are hospitals, the cross health centers, the skier...* is frankly dangerous, because you can go back to your computer after a tiring week of field surveys to discover that it doesn't recognize them and it places another icon in its place by default. This is even more important if you're going to share information with others or use various programs that aren't from the same company.

A useful illustration of these warnings is presented with the image below. After the 2004 Tsunami in Muelaboh (Indonesia), we took points of a water network and drew diagrams of the connection and diameters of pipes as remembered by the plumbers in an effort to reconstruct a network plan lost to the waters. Once situated in an aerial photo you could see where each pipe went and reconstruct the plans.

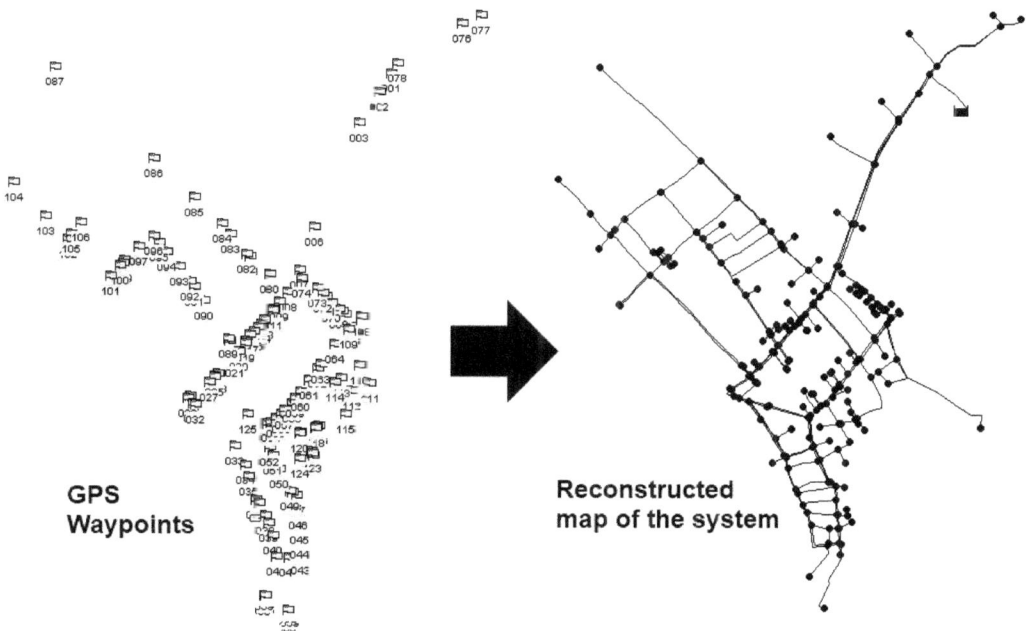

GPS Waypoints

Reconstructed map of the system

To come up and write the name of a hundred odd points would have required a lot of invention and would have put anyone's patience to the test... Furthermore, the tags would have made the visualization too crowded. Simply following the numeration proposed by the GPS is much faster and it permits you to better orientate yourself because you know which points you took and in what order.

Also observe that if you had used icons to distinguish between the Ts, elbows, end of line valves and the computer didn't respect them, all your work would have been useless.

Instead the GPS simply went along taking points while in the notebook the connection scheme was noted by the GPS number:

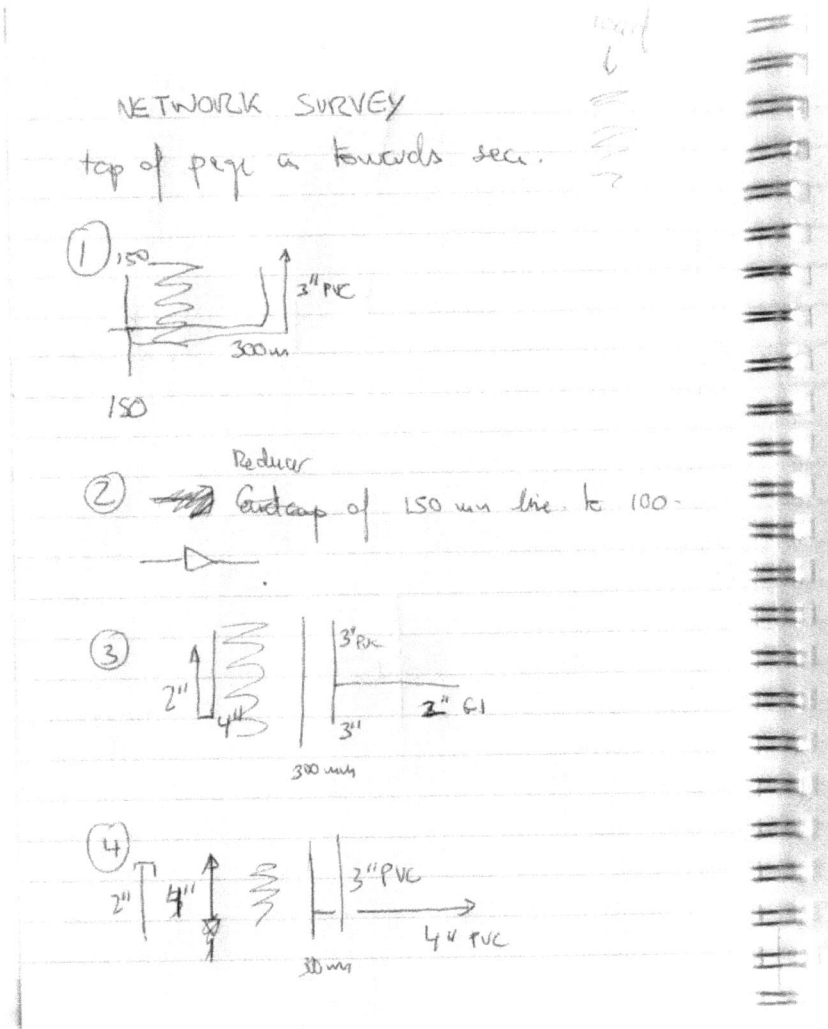

Traces or Tracks: Which path have we followed?

In a city it's easy to see that the piping follows the streets, in a rural project it's not so easy.

When a GPS is turned on, it automatically saves the track it's been on, like Hansel and Gretel and their breadcrumbs trail. The traces are successive points taken every now and then. This lets you trace the route answering questions of a type such as which path have we followed... (this morning, while driving on the road, while doing the health survey...)?

The image shows a blue trace of the track created by the GPS of a smart phone after a walk.

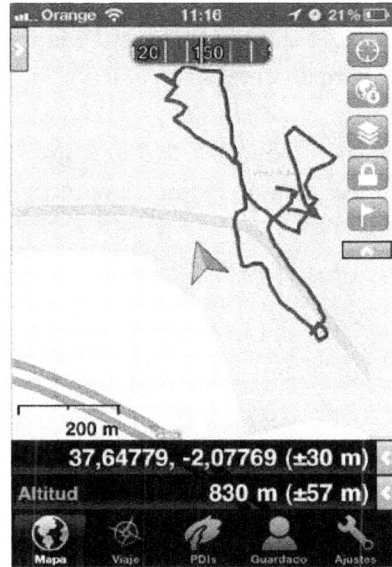

Other interesting uses:

- Closing the route also allows you to **outline plots.**
- It permits you to find out **real distances** of routes, and not just straight lines distances.
- When tracing a path you can trace its **aproximate topographic profile** on the internet, as you've seen in excersice 4.

The GPS has a limited memory. Although usually it can save all the points you might need, with the tracks it becomes full easily. Once the memory is full, they tend to erase the old with the new, in other words, the end of the route erases the start.

The routes: where to...?

The routes allow you to plan a track on a computer. You can mark the points that make up the route in a computer and them follow the directions of the GPS in the field.

Even though they aren't very useful for mapping, because they start from something already mapped out, they can be very useful to plan routes.

An interesting application is the planning and following of **transects**. A transect is a way of observing and collecting data following a predetermined track. In its simplest form, a line is traced and it is followed while observing some kind of phenomenom, for example, defecation in the open air. Following a line has various advantages:

- You can find out **gradients** and predominant directions, in other words, if something occurs more or less advancing in one direction.

- They can very easilly calculate **densities** (of tress, open defecation, etc.).

- They increase the **impartiality and objectivity**, avoiding taking data in a biased or senseles way due to you not knowing the terrain, being taken/ taking yourself to what you/ somebody else wants you to see, or simply asking at the side of the roads because it is more convenient.

Be they points, routes, tracks or areas, the GPS works them up using coordinates. In the next section you can see the two coordinates systems most frequently used.

Projections and coordinates

On a flat map, you can't accurately represent areas, angles and distances all at the same time. One of the three distorts itself in the process of *flattening out the Earth* to represent it on paper. The different ways of flattening out the Earth are called **projections**.

What can be done is deciding which of the three variables will be distorted and to what extent, depending on the projection that is used. For example, the more local the projection, the less the distortion due to there being less *wrinkles to iron out*.

Geographical coordinates (latitude/longitude)

They look like this:

48° 03' 38" N   254° 21' 55" E

The origin of these coordinates is the centre of the Earth. Imagine for a moment that there was a canon placed there and that, instead of giving the position of a point, we want to fire at it. We can say, swing the canon 254° clockwise, and raise it 48°.

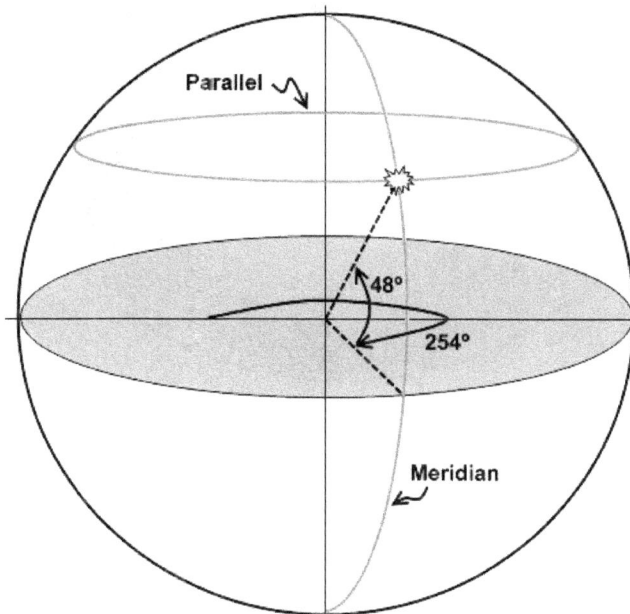

Latitude is the vertical angle from the equator and **longitude** is the horizontal angle from an arbitrary 0° prime meridian, the meridian of Greenwich. Therefore, there is north or south latitude and east or west longitude. Since it is shorter from the west, 254° is usually expressed as 106° west (360-254= 106° W). Another way of expressing it is simply as -106°.

The fundamental disadvantage of the latitude and longitude is that it uses degrees, minutes and seconds, so that 60 seconds form a minute and 60 minutes a degree. It is not very straighforward to use divisions of 60 to take measurements. Thus, although you can observe roughly where it is, it is difficult to determine the precise coordinate of the point.

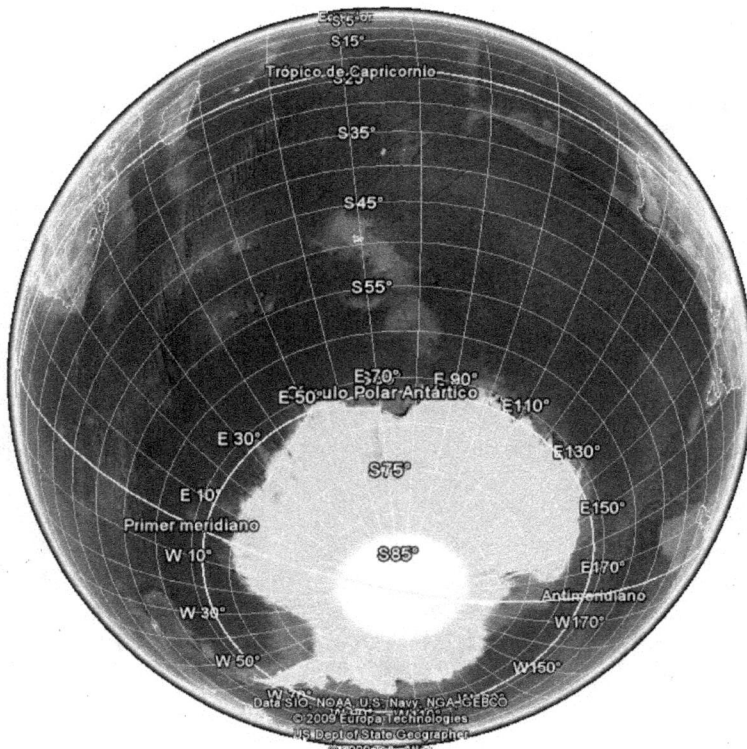

UTM Coordinates

They look like this:

26E 357641 9164532

The UTM system of coordinates is based on the UTM projection - the points are projected onto a cylinder that surrounds the land. That cylinder is then unrolled to obtain the map.

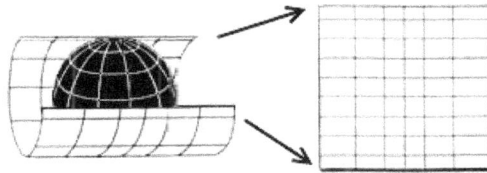

UTM projection

The grid is constructed dividing 60 vertical zones and 20 horizontal bands. The zones are numbered and the bands are assigned letters. In the coordinate *38K 452491 7655296,* the point in question is found in the zone 38 and the band *K*, a rectangle over Madagascar:

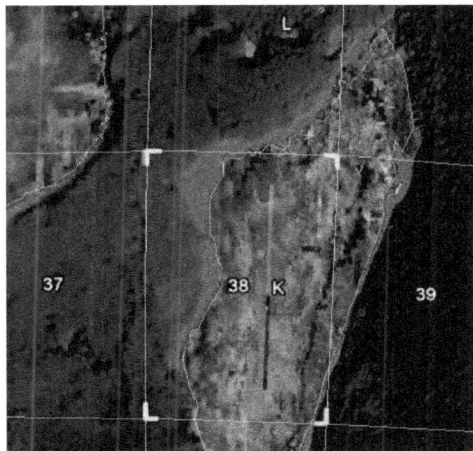

For the whole globe the division would be:

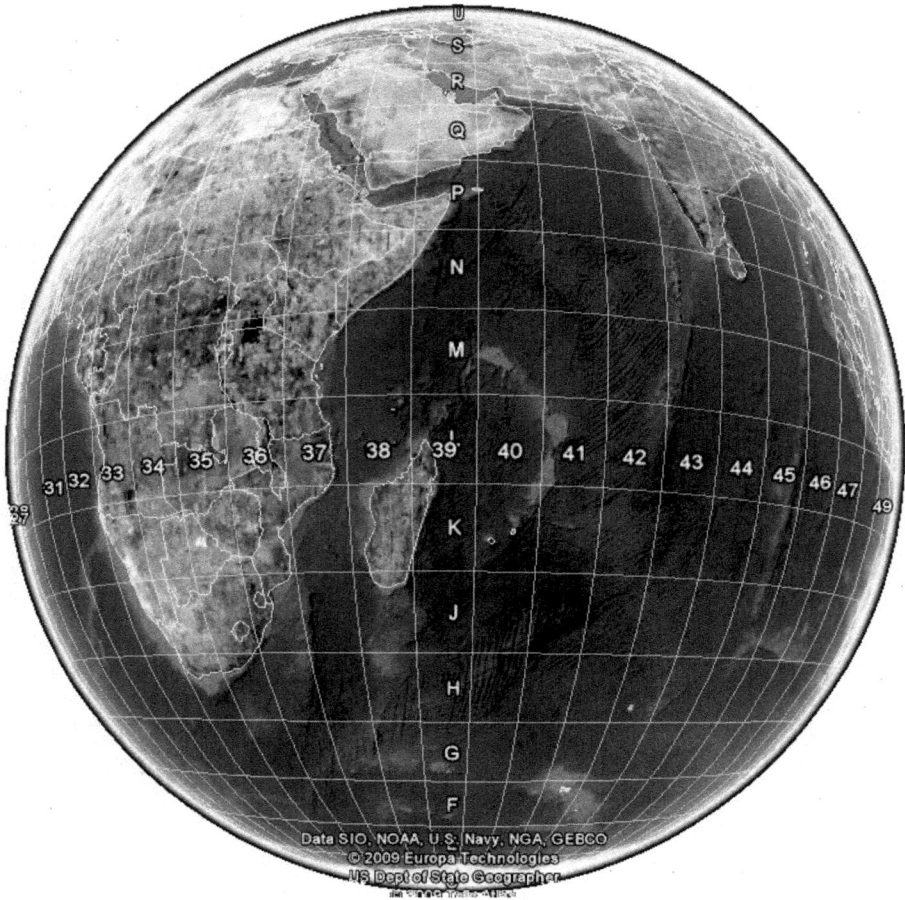

The two groups of numbers that follow are the horizontal coordinate in meters (452491) from an arbitrary point called the False East, and the vertical (*7655296*), also in meters, this time from the False North (they´re false because they are only a convention to avoid negative coordinates). Since all express distances in meters, it is very easy to determine the distance between two points by subtracting their coordinates. The point *38K 452000 7655000,* is 491 meters horizontal and 296 meters vertical from the previous one:

$$
\begin{array}{r}
38K\ 452491\ 7655296 \\
-\ 38K\ 452000\ 7655000 \\
\hline
491 \qquad 296
\end{array}
$$

Navigating with coordinates

When a map has the UTM grid printed on it, it is very easy to locate a coordinate or to read the coordinate of a point on the map. Take a look at this map:

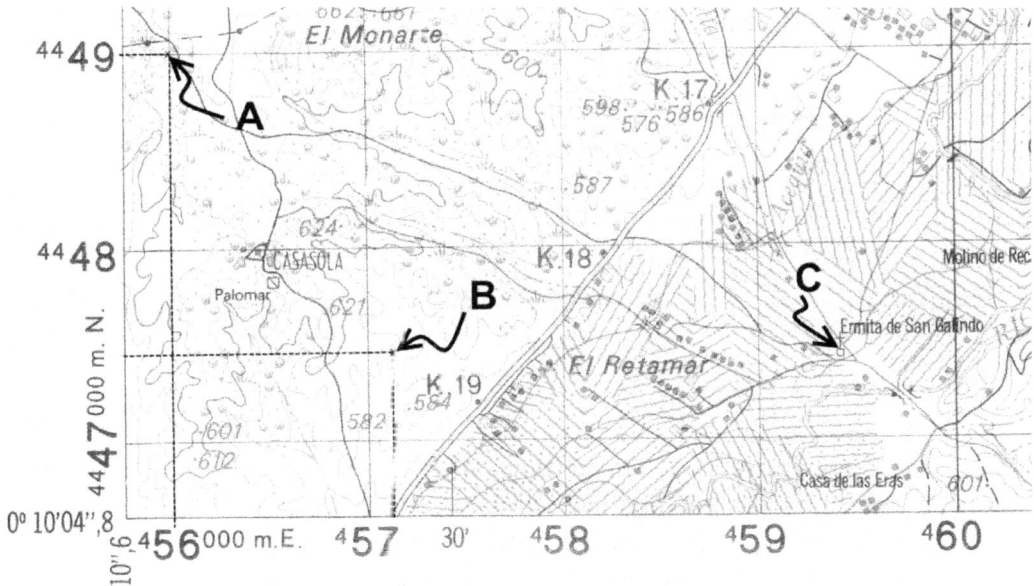

The numbers ^{4}56, ^{4}57, ^{4}58 etc. correspond to the coordinates 456000, 457000 and so on in which 3 zeros have been eliminated for convenience. The same occurs with the verticals. The grid has 1000 m sideways (457000 − 456000 = 1000). Since the band and zone will be the same for all these points, we can disregard them, we can see that:

- **Point A** coincides with the intersection of the UTM grid at the lines ^{4}56 y ^{44}99. It has the coordinates 456000 4449000.

- **Point B** is approximately 1/10th of the horizontal distance between the lines ^{4}57 and ^{4}58; as 1/10th of 1000 is 100 m, its horizontal coordinate is 100 m added to ^{4}57, hence 457100. The vertical is a little less than halfway between the lines ^{44}47 and ^{44}48; its vertical coordinate therefore is 4447450.

- What is the coordinate of the Ermita de San Balindo (point C)?
  Solution: 459400 4447400

Sources of error and precision

1. The relative position of the satellites

The number of satellites important but aso where they are, so that the intersection between their signals is as clean as possible:

When the satellites are very vertical or somewhat aligned (A) the intersection is extended along a larger area and the position is determined with less precision. The most clean intersections occur with satellites that are lower and further apart(B). The problem is that these satellites whose signal you loose more easily.

Estimation: The GPS will give you an estimate like "Ready to Navigate. Accuracy 9m". Some also have a value of DOP[1] a more precise way of expressing precision that also allows you to compare different DOP values. When possible, try to work with DOP values below 2 and in all cases below 5. For marking points that don't need great precision, for example locating a town, you could use higher values (although it's simpler to take a different point in the town with better visibility).

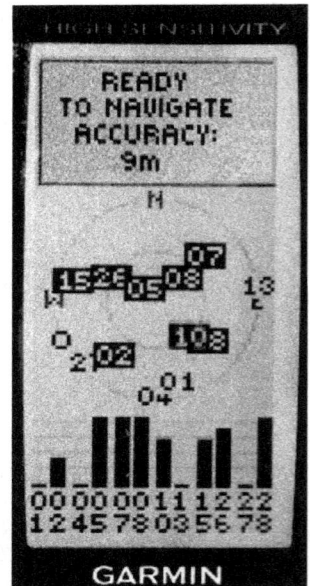

The majority of GPS show a graphic, like this "Garmin Vista" that combines the quality of the signal (height in the bar) with the relative position of the satellites, numbered from 1 to 24. You would be situated in the center; the exterior circumference is the horizon (satellites with best intersection!) and the smallest circle corresponds to an angle of 45° with the horizontal.

---

[1] *Dilution of Precision*

Correcting measures:

- Move the GPS away from the body (avoid the heads of curious people!) and elevate it like a person who raises his hand to ask questions, to avoid putting yourself between the GPS and the signal.

- Average out the points. Some GPS, instead of taking a point, allow you to take lots in the same place and then find the average. The Magellan receivers do it automatically if they don't detect movement.

- Avoid the more humid hours in places with a lot of vegetation since water reduces the signal.

- In abrupt places that make the signal reception difficult (valleys, urban canyons) use an exterior antenna.

- Move to take the point if there is a place nearby with better signal.

2. Multi-lane error

The GPS signal bounces off rocky walls, trees, buildings, interiors, etc. When the signal bounces it takes a longer road and the GPS *thinks* that you are further away.

Correcting measures:

- Whenever possible, look for clear places in open areas. Although we can't always choose the places where we need to take a point, we can avoid situations that cause multiple re-bound, for example, by putting the GPS in the interior of vehicles.

3. Ionosphere error

The ionosphere is a layer of the atmosphere full of charged particles that get in the way of the GPS signal, diminishing its transition speed. The quantity of the ionosphere that the signal traverses depends on its inclination angle (A vs. B). The lower, the more ionosphere it will have to go through and the more the signal is affected. In a certain way, and to help you remember it, it's similar to how the atmosphere affects the lower rays of sun creating amazing bigger sunsets with orangey colors.

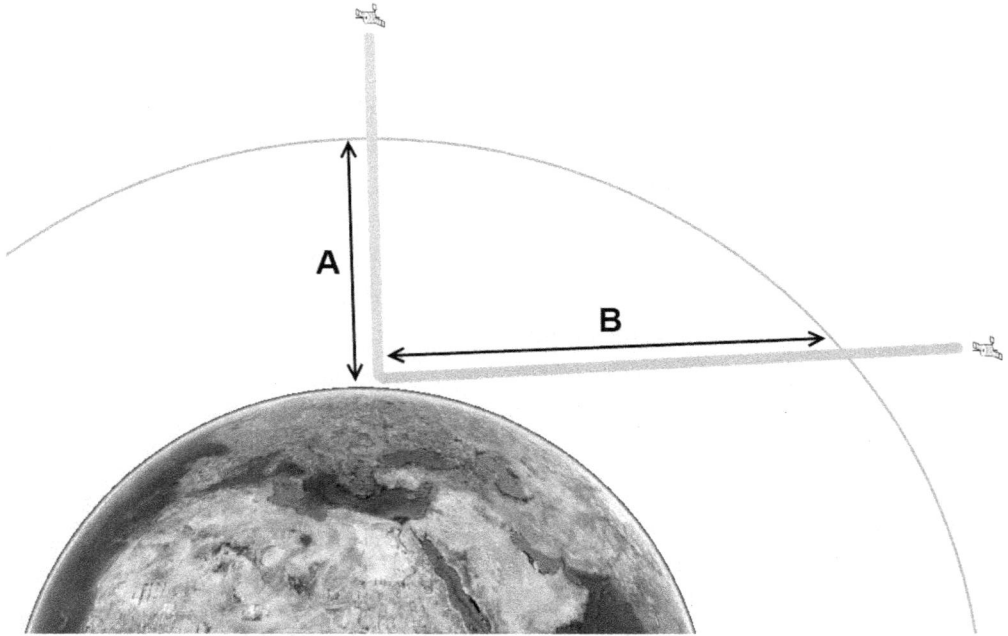

Correcting measures:

- Acquiring more expensive models is rarely worth it for standard applications. These models can calculate the influence of the effect, measuring the time lag that occurs when two signals have left the satellite at the same time but get delayed to a different degree due to their frequency difference.

- WAAS, EGNOS and MSAS are the North American, European and Japanese versions of the same thing: an earth stations network to calibrate the GPS signal and increase its accuracy for aviation use. Some receivers have the option to activate or deactivate it. If you have this option, deactivate it. This correction isn't available in the major part of the planet, especially in those areas where Development work is concentrated, and in this way you will save on batteries.

4. Intentional error (SA)

Up to May 2000 the State Defense Department of the EU, that administers the system, intentionally downgraded it to avoid hostile use. Even though this downgrade isn't used generally anymore, if you're in active conflict zones you can find the signal that affects that specific zone is downgraded. In a conflict area, be very careful about using the GPS without having the authorization from all sides involved!

Cold starting a GPS

In addition to the position, the GPS signal transmits two other things that help their work:

- The **almanac** contains the orbital information of the satellites to allow the GPS to know where it has to look for the satellites according to the time and the day. It takes around 12 minutes to download and it is no longer accurate after 3 or 4 weeks.

- The **ephemerides** contain more precise infromation about the orbits and other data like the clock error. They take 30 seconds to download.

When you start a GPS for the first time, you've changed position hundreds of kilometres or its more than a month since it's been turned on, it doesn't have either almanac or ephemerides. In this case the GPS **starts from cold**. Even though it can start to locate satellites and give positions before 12 minutes, it's convenient to wait this time so as not to loose the signal during the route and to raise the precision.

If the GPS already has an almanac that's up to date, you will still have to wait 30 seconds if it's missing the ephemerides (**warm start**). If it has both, you can start straight away (**hot start**)

If the GPS is going to cold start, try turning it on quite a while before getting to the place of work. For example, you can turn it on and leave it on the dashboard whilst you travel:

An indispensable ritual: The GPS configuration

When using a GPS and interchanging information with other people or devices, like your computer, make sure that:

1. You're using the correct **datum**.
2. You're using the same **coordinate system**.

1. The datum

Imagine that you have a coordinate, for example, 220 km south and 321 km east. As it is it's not very useful, you need to know where to measure from! The same applies to the GPS, it needs to know where to measure from.

The **datum** is used to make sense of the coordinates. It's a combination of the projection that is used and some point or points of reference from which we take the measurements. For example, if we *assume the Earth is a sphere and we are measuring from the North Pole.*

The important thing **is that the GPS and the map you use with it have the same datum**, so that one isn't measuring from the North Pole and assuming that the Earth is flat and the other from Bishops Itchington (Staffs, UK) supposing that the Earth is spherical. Also if you're going to collaborate with others, you need to **agree on the datum** you're going to use.

Usually the choice is between two options, either the universal datum **WGS84** is used (remember this name!) making the exchange between collaborators easier and being compatible with generic services like Google Earth, or a local datum is used that increases accuracy. **If you are going to work with Google Earth you have to use WGS84.**

The data can be easily changed from one datum to another if necessary. The majority of GPS programs and the GPS itself will do it. The important thing, **is that you don't take data thinking that it is referred to one particular datum when in reality it is referred to another.**

If you already have points you can ascertain its datum in these ways:

- If the file has an GPX extention, that is generally used for GPS data exchange, the data is always WGS84.

- In other types of file, the datum comes included as part of the heading of the file. For example, in this WPT file, it is WGS84:

The majority of GPS file types can be opened with the notebook or with a spreadsheet like Excel.

To ascertain the datum of a map read the legend if it is printed or a digitalized print, or try to find the metadata (data about the data) if it is digital.

Sometimes, the datum isn't clearly identified as such in maps. This is the case in this map of the Atlas Mountains in Morocco, but you can guess that is Merchich, because it will be amongst the list that you have in the GPS:

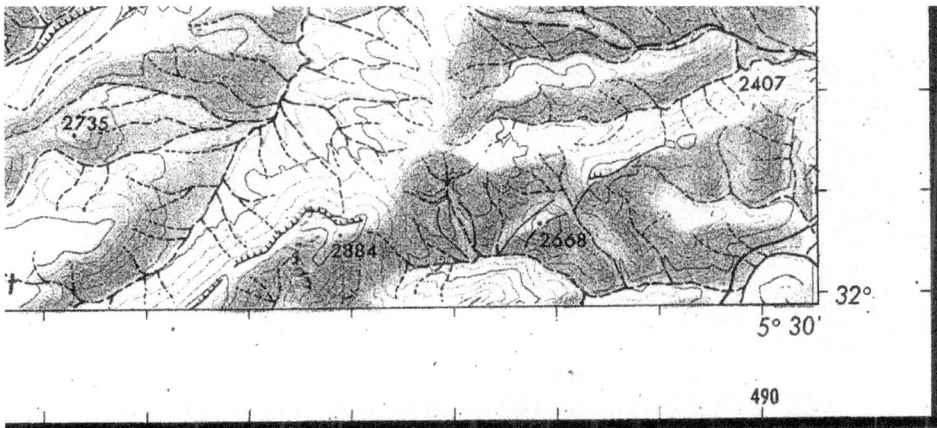

Ellipsoïde de Clarke 1880. Système géodésique Merchich.
Projection conique conforme de Lambert Nord-Maroc.

2. The coordinates

The type of coordinates that you are going to use should be those used by the people or the programs with whom or which you are going to work.

Unless there are other requirements, always try to work with UTM. The coordinates of longitude and latitude were invented so as to navigate without obstacles. They are useful for both sea and air. On land we cannot determine the distances between two points or the position on a map without laborious calculations. But above all, they are liable to error if the people note the coordinates by hand (something to scrupulously avoid!) on a form. Note, for example, the importance of the comma in these 3 ways of expressing the measurements:

-78.1947°	Degrees and decimals of a degree.
-78° 19.47'	Degrees, minutes and decimals of a minute.
-78° 19' 47"	Degrees, minutes and seconds.

These three points are very distant, up to 14.5 km between them. As you can see in the image of Central America, the error is considerable. If you are not scrupulous with the comma you could end up after your survey with the points being in the Indian Ocean or in the neighboring country.

Despite all this, the geographical coordinates are widespread and frequently they are installed by default in programs and GPS devices, which is why you will probably end up working with them. **GPSvisualizer**, which you will use for the exercises, **requires decimal degrees.**

The configuration ritual

Now that you know what everything is, how to find it out and which is more convenient, this is the procedure to follow to configure your GPS device. It may vary according to each brand, but normally you will have to go to the Settings Menu on your GPS:

1. **Configure the datum by** selecting it from the list of possibilities. In this example, we will select WGS84 because we intend to work with Google Earth and GPSvisualizer.

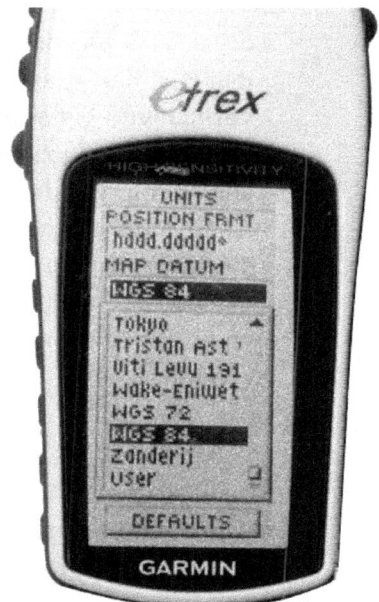

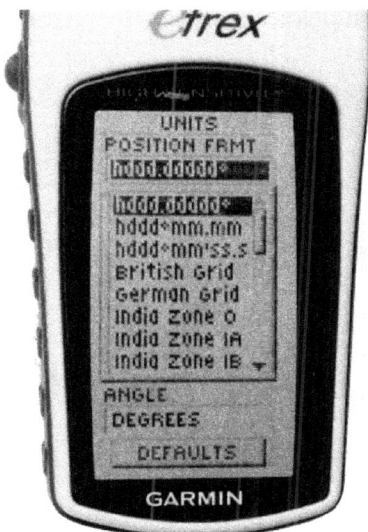

2. **Configure the coordinates** in which you are going to work. In the image, they have selected degrees with 5 decimals, that is to say, 41.57367º.

That is all you have to do.

Working with a GPS

Once the GPS is switched on and you have the almanac and the ephemerides up-to-date, it is ready to work.

Ergonomics to improve the signal:

Most of the time the signal is good and its not so important to follow these steps. Nonetheless, it is no bad thing to get into the habit of doing this to improve the performance of the system:

- If the GPS antenna is **helicoidal** (A), that is, it has a nozzle, carry the GPS **vertically** to better receive the signal. Nevertheless, this antenna is less sensitive to the device being inclined. If you don't see the antenna externally then it probably has a flat one (B); the manual will confirm it. In this case carry the machine **parallel to the ground**.

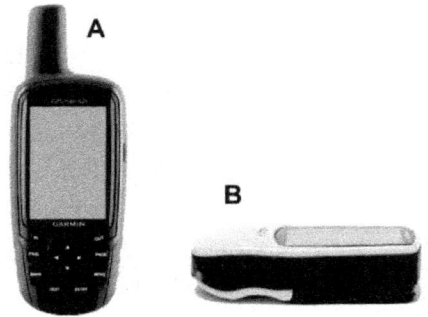

- Distance the GPS from your body as much as possible; hold it with your arm extended. Evidently if you have a flat antenna, you will have to incline the GPS so that you can read it, but try to take the point with the GPS horizontal.

- It won't always be practical to carry the GPS in your hand all the time. Often you can carry it in a backpack, as long as no metalic objects cover the antenna, or inside a car without major problems. The dashboard would be the best place to avoid rebound with the roof and have a better signal. Carrying it around your neck is not a good idea as you block out half the field of reception, the antenna isn't horizontal and every time you bend down it will perform a Tarzan swing against something. If you are frightened of losing the signal, a good alternative is to fix the GPS with velcro to a backpack at shoulder height or to fix it on the body of the car.

Managing the batteries:

They normally use standard batteries, AA type, that should last between 6 and 12 hours uninterrupted, sometimes more, depending on the use that they are given. In whatever case, here are some points to keep in mind so that they last longer:

- **Disactivate** the **WAAS/EGNOS** option.

- The majority of GPS have a **save battery mode** in which they take the signal every 4 or 5 seconds rather than all the time. If the signal is good and the work day is long, or simply to avoid creating too much waste in a country where no adequate waste management systems exist, perhaps you could activate it.

- **Avoid rechargable batteries**. If the previous point makes you think about using rechargable batteries, think again. In our experience, rechargable batteries have a tendency to run down, deteriorate and make you mistake flat batteries for charged batteries. How can you trust them in a job that you have no fervent wish to repeat?

- Always **carry spare batteries** and avoid using poor quality batteries or those that have been stored for a long time. You do not want to put up with a bumpy

500km in an off-road vehicle only to discover that your Dragon Power batteries are little more than wrapping.

- They are affected by the **cold**. Carry the GPS close to your body and inside your clothes when it is very cold and they will work better and last longer. Warm the batteries by putting them next to a warm part of your body for several minutes. By warming the batteries against your body, you will be able to take some last vital points before they completely die. Never heat them in any other way, they explode!

- Even though its obvious, perhaps its only necessary to turn the device on, take the point and then switch it off. In these conditions, the batteries last several weeks.

Your safety:

If you want to stay in one piece or simply avoid having an awful time, for your security and that of your colleagues, don't ever forget these two pieces of advice:

- Never use a GPS in a **sensitive zone**, near a military base or in areas of conflict without obtaining the **goodwill** of all parties. A GPS, like a camera, gives raise to all sorts of suspicions. Don't forget, safe use!

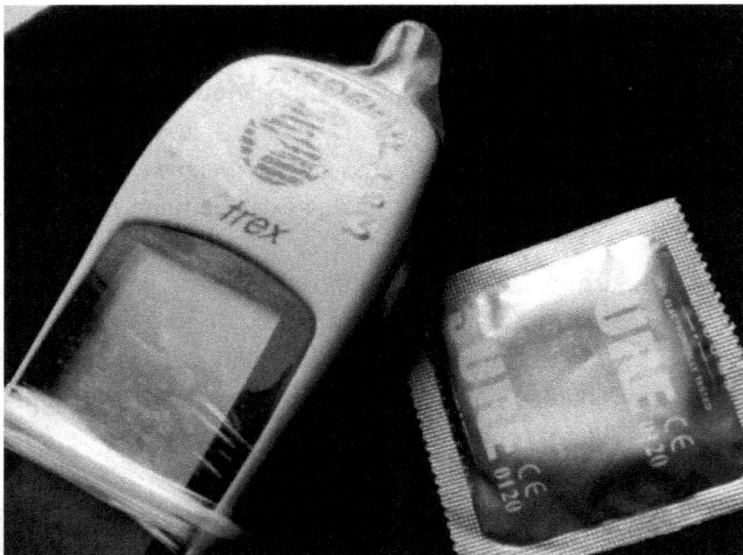

- Some places require the use of a GPS to **navigate** and not only to work. Such situations are clear, like when working in desert areas, jungle, mountains, foggy and variable weather areas … Others are not so clear. Even when you have local people that know the area:

  - Always take **reference points** in key locations: before setting off, in the last habitable place, where you have left the car, and so on. In these cases, don't trust everything to your GPS: carry maps, a compass and know how to use them.

  - Register tracks by keeping the GPS on. The GPS has a **trackback** option that allows you to reverse the path you have walked.

Saving the data:

Some quick suggestions about taking data:

- Imagine that you have done a survey of 274 points. It is highly likely that for some of these you will forget to mark the point or that your finger suddenly swells and you press the wrong buttons. With luck you will realize that one is missing, but which one is it?! You can easily lose all your work like this. **Every 10 points, take the same point twice**. In this way, if at some point you don't mark one, you can notice that one point is missing in the interval between this double point and the previous. For example, if the O are the double points, the third interval has several points missing:

  OOxxxxxxxxxxOOxxxxxxxxxxOOxxxxxxOOxxxxxxxxxxOO

  This ritual also keeps you alert to the numeration displayed in the GPS.

- One of the quickest ways to work is **to take a photo of the notebook** with the name of the place and the number of the waypoint. After, take all the photos of the place. When loading all the photos on the computer, they arrange themselves automatically in the order in which they were taken, allowing you to know which photos correspond to which location without working any further. Another of the advantages of this method is that you digitalize your notebook; it's so easy to leave it somewhere and lose those important notes!

  In the image, for example, the notebook has been photographed in the course of an evaluation of the economic/food situation in Somalia for a family in Xarxar village. This survey, (the answers to the questions are numbered) corresponds to the point GPS 65:

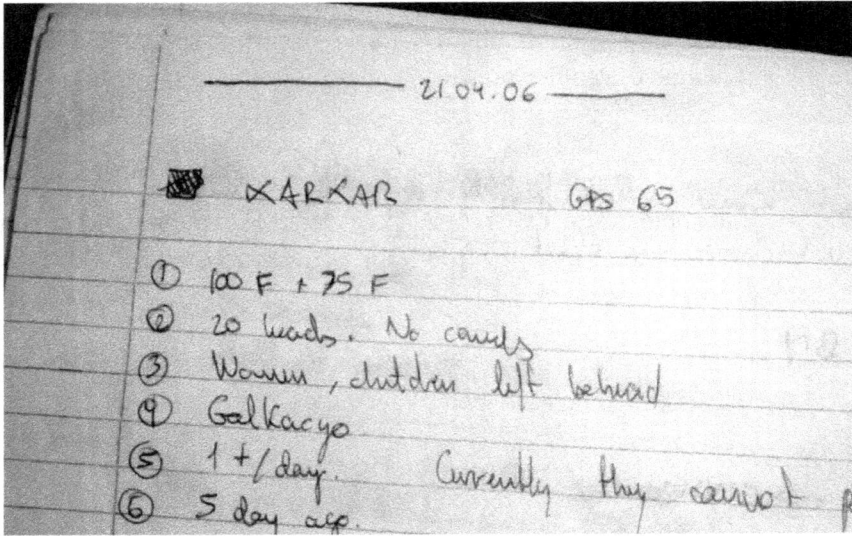

Each point (63, 64, 65…) and the corresponding photos are ordered automatically:

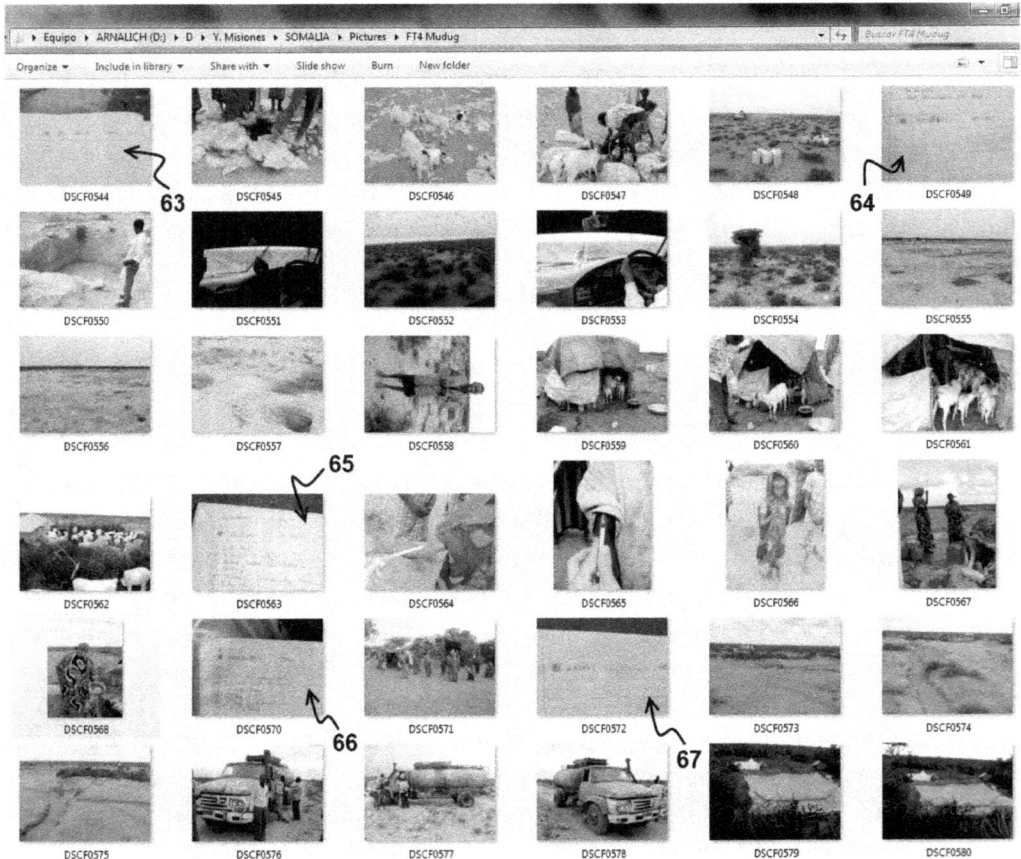

In places without an actual name, you can photograph the GPS with the coordinate and later the area so that you know that the images correspond to each of the waypoints. In the example, the ravages of a Tsunami at different points:

IMG_0026   IMG_0025   IMG_0024   IMG_0023   IMG_0022

IMG_0019   IMG_0018   IMG_0017   IMG_0015   IMG_0012

- **Begin and end each track with a GPS point** that you register. In this way you can easily recognize it and also name it, for example, track "68-121".

- In many GPS, there is no way to avoid registering tracks if the GPS is switched on. Also, many have a function called **dead reckoning**, this means that if it looses the signal while tracing a track it will continue to trace it with the direction and speed you had until it receives the signal again. The result is that you end up with a confusion of tracks and don't know which is which. **Save the track when you finish** so that you don't add things that you don't need.

- Remember that the GPS has limited memory. **Check the percentage** you have used every so often to avoid accidental overwritting.

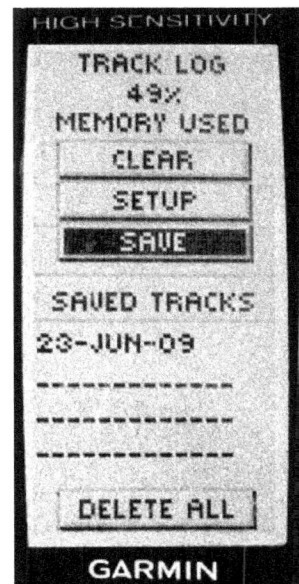

HIGH SENSITIVITY
TRACK LOG
49%
MEMORY USED
CLEAR
SETUP
SAVE
SAVED TRACKS
23-JUN-09

DELETE ALL
GARMIN

Uploading and downloading data

Once the data has been taken you need to transfer it to a computer so that it can actually be useful. The proccess is very similar to that of connecting a digital camara to a computer.

Never copy coordinates by hand, even though a surprising number of forms ask you to do just so. Each coordinate has some 20 digits. If you copy them by hand to a paper and then to a computer, its very probable that you will introduce errors that are difficult to check or detect.

To avoid copying them by hand, you need some of these things:

- A **connection cable** to the computer.

- A **program** or **website** that allows access to the GPS and uploads and downloads data to your computer or the cloud.

The proccess should be simple enough although depending on your computer and the GPS device it can sometimes be very time-consuming to set up.

The connection cable

Some GPS devices come with them and with others you have to buy cables separately. Just make sure that you have one. You may be in one of these two situations:

- **The cable is USB**. For the majority of GPS devices you don't have to do anything else, it is a USB cable like the one you would use for a digital camara. This is the most common mini-USB connection:

- **It is something else**. For the older or simpler models this cable connects to a Series 232 o Mini-DIN port, ports that the majority of computers don't have anymore. In these cases, you need a USB-Series or USB-Mini-DIN adaptor, another cable that complicates things a little. This adaptor cable comes with a CD and needs the controller to be installed so that it works properly. In these cases the connection would be:

GPS Cable   Adapter

The programs

There are numerous GPS programs that allow you to download the points and tracks from a GPS or load them to a GPS from your computer. Google Earth at the time of writing is unbeatable if what you want is to interact in real time and place the points on a map. However, Google Earth has its downsides: obtaining a file with coordinates is not direct, it is not straighforward to work with other maps or aerial photos that aren't its own, it doesn't support many GPS models and it fails too often.

If you use Google Earth:

1. Connect and switch the GPS on.

2. Go to > Tools / GPS and fill in the dialogue that opens depending on the model that you have:

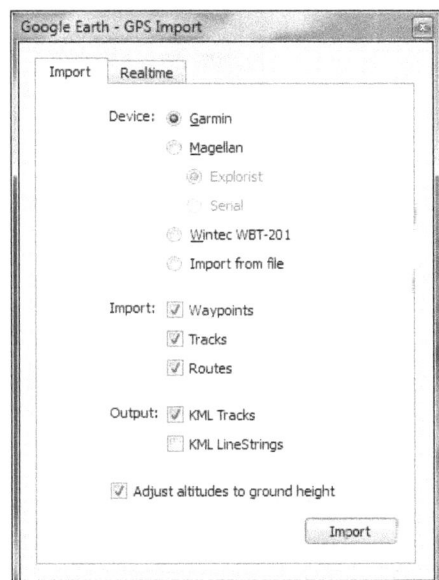

Note that there is a tab Real Time. This tab allows you to show your actual position on the map and to see how it changes as you move. The epigraph "Navigating with Google Earth in real time" explains this later.

If you prefer to use other programs:

You can find below links for some of the other most popular programs:

- Oziexplorer: www.oziexplorer.com
- CompeGPS: www.compegps.com
- Gpstowin : www.gpsinformation.org/ronh/g7towin.htm
- EasyGPS: www.easygps.com

No matter what program you end up using, the crucial requirement is that your computer detects the GPS and can communicate with it. In the majority of cases it is automatic, in others you have to try out the ports.

This process is done with Oziexplorer in the example below. The steps will be similar in in other programs:

1. Go to Configuration and click the GPS tab. There, introduce the model and make of the GPS device and click on Find GPS so that it tries to detect it automatically. Don't forget to have the GPS switched on!

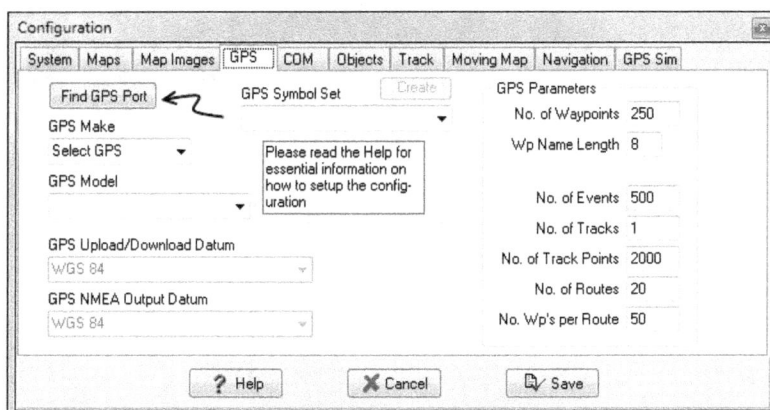

If all goes well, it will respond with the port number to which the GPS is connected (COM15). It is not a bad idea to try and remember it for future uses.

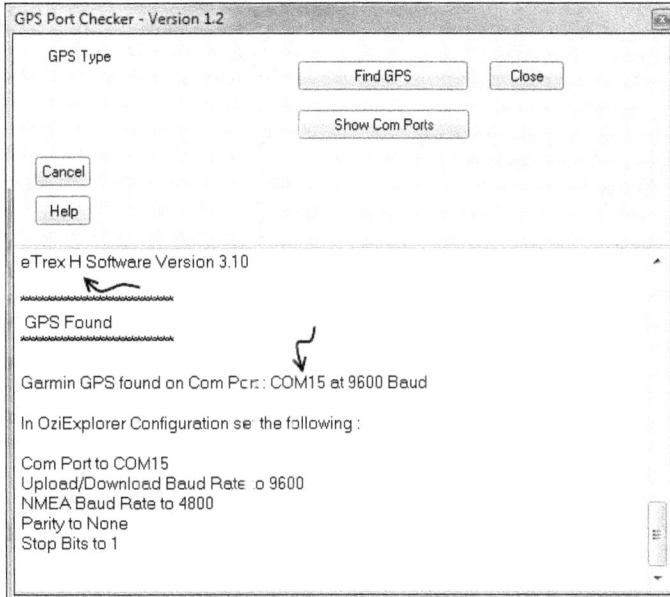

2. If things don't go well, then you can start a process of various adjustments. It is a trial and error process and therefore dificult to explain here, however here is a guide of things to be tried in this order:

2.1 Begin by checking that the GPS is switched on and the cables well connected… hmmmm, check it!

2.2 See if the GPS is detected when using a website instead of a program (next section).

2.3 Have you installed the adaptor cable driver if you are using one?

2.4 Check that there are no other drivers to install.

2.5 Some GPS devices have "Force Send" options, check it.

2.6 Try configuring the connection port manually, checking one after the other:

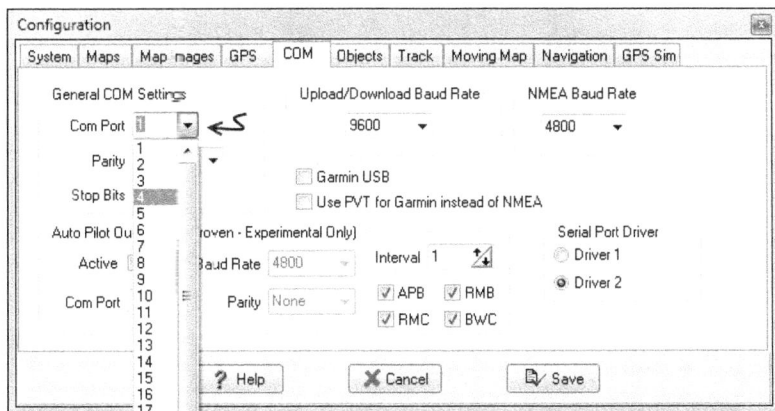

If all goes well, you can now use the program with the GPS. In this case, you have to search in the menus for options such as "Download points from the GPS".

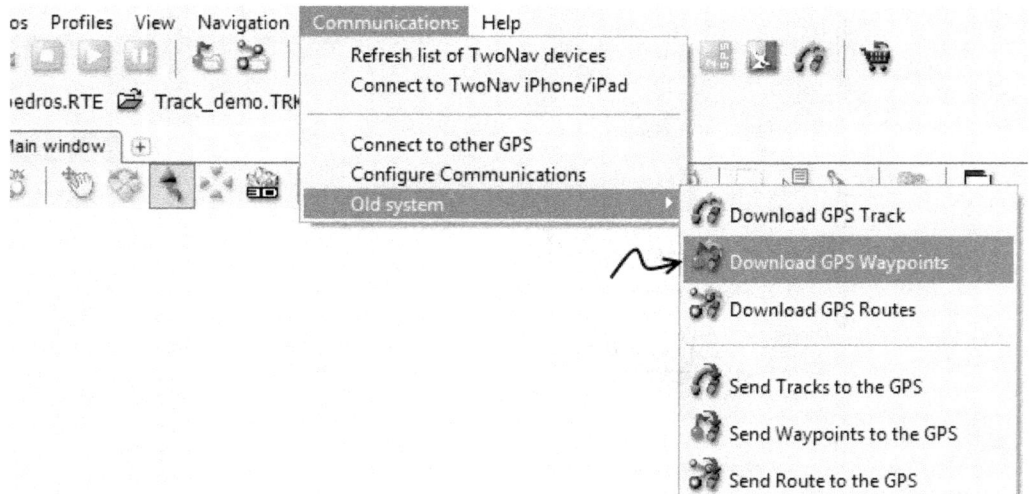

Take advantage of this opportunity and study the rest of the menus. In these programs the download, upload and configurations settings are grouped together under a menu called GPS or Communications.

Online

Many manufacturers already have a website that downloads and stores the GPS data and allows you to download it in various formats. Below we show some screens from the Garmin manufacturer's site:

http://connect.garmin.com

1. You must create a user account and log in so that they remember that the routes and points are yours. Then click Upload:

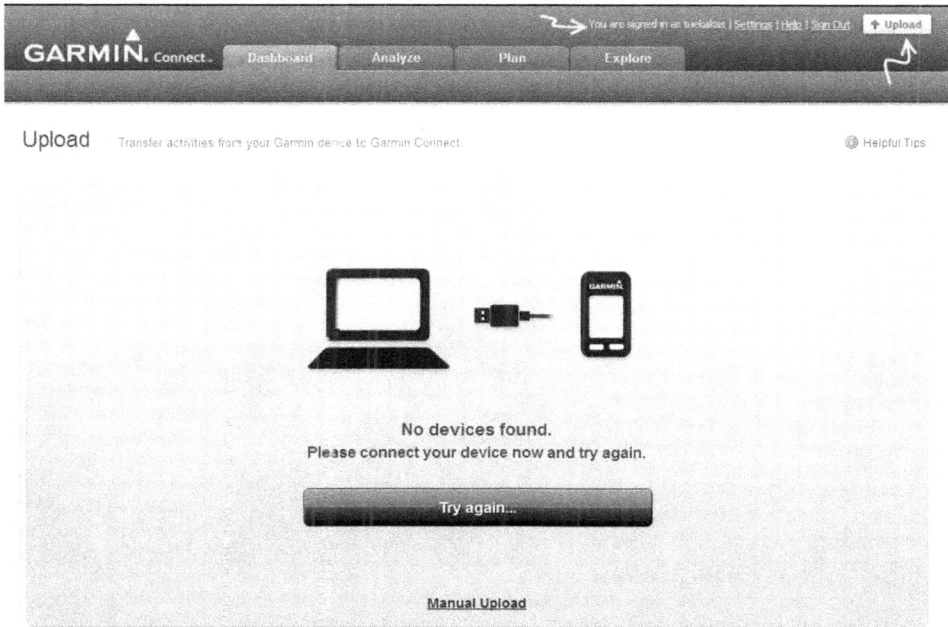

2. Connect the GPS switched on and wait for it to be detected. When this happens, select Upload New Activities:

3. In the list of activities (as it is sports-orientated web interface, surveys are called activities), select the one that you want to download. This activity, for example, corresponds to the proposed route for pipeline layout in Haiti:

Formats

There are many differen formats for GPS data: GPX, WPT, KML…. Depending on the model you have and the programs you use you will have one or another. **GPX** is the format for universal exchange.

To convert some formats to others you can use a website or download the **GPSBabel** program: www.gpsbabel.org. With this, you only have to choose the input formats and the Output that you want, navigate to the file origin and give a name to the new file destination:

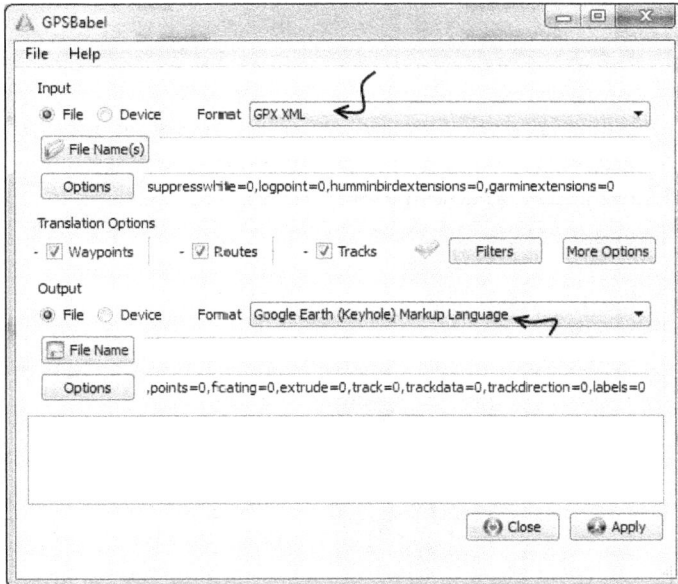

Navigating with Google Earth in real time

Navigating in *real time*, you can see your position and the path that you are following at every moment in Google Earth (as well as in other programs). The GPS already does this but by connecting it to another device you can see your position on satelite images. You can connect the GPS to a computer, a PDA or have everything integrated in a smart phone.

Navigating in real time is very useful in cases when you cannot wait to see the information later, because you have to make decisions right then or because it is a complicated case. Even though the telephone option is very convenient, the information often ends up squashed together in a very small screen, and it may be better to opt for a computer.

Normally, you will need to have telephone access to the internet with an actual phone or a USB modem (A).

With the GPS connected to the computer and switched on (!) you must go to Tools / GPS and select the tab Real Time. Choose the protocol type of your GPS, check the box Automatically follow the path and click Start:

Using Google Earth offline

You don't need to have an internet connection all the time to use Google Earth, you could have previously saved the areas that interest you in the cache memory by following this process:

1. To access this option go to Tools / Options and, in the dialogue that opens, select the Cache tab. Select the maximum, 2000:

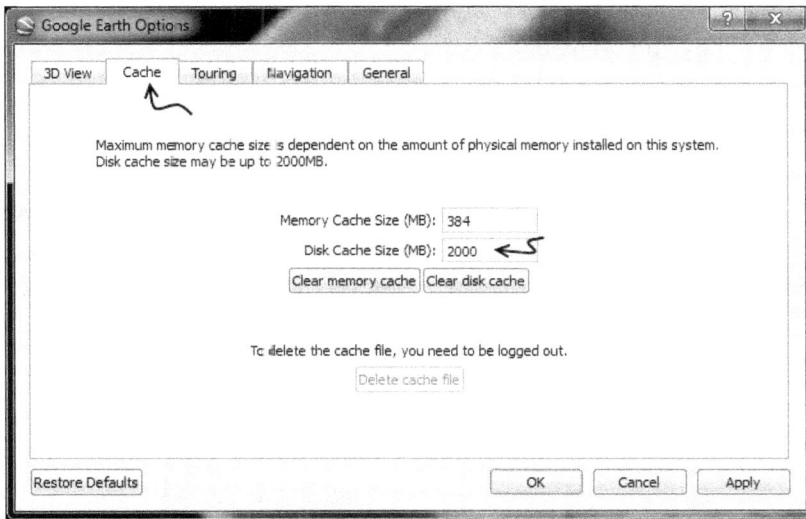

2. Delete the memory cache to ensure that you get to use the maximum capacity and apply the changes before leaving the menu.

3. Explore the area that interests you in the detail that you need while you have internet access. If you want to have some layers such as Borders, Highways etc, tick these layers:

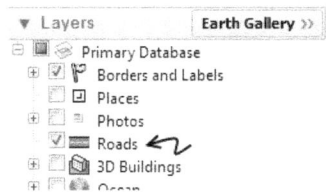

The cache memory is recorded in several files that include the word cache:

If you need more memory you can carry on saving copies of these files within folders named after the areas covered. To change the area, copy the files of the place's folder and overwrite the existing ones in the Google folder.

The route to these files changes according to the operating system that you have, it is probably best for the system to search for them using the computers search option for the C: drive.

There are programs that can scan areas automatically for you (i.e. Google Earth Cache Builder, Google Earth Voyager, etc.) Search the internet for the later versions to determine if they can be helpful to you.

Usually the cache files that you prepared on a computer can't be sent to another computer, unless the file routes are exactly the same, which is both unlikely and impractical. Before going to an area without internet, disconnect from the internet and test Google Earth to make sure the memory is working as it is meant to.

Using third party maps

Sometimes, the Google Earth maps aren't practical or they don't have the information you need. In those cases you have various alternatives.

1. Using the GPS with a printed map.

When you're in dangerous territory and you don't want to rely on electronic devices it is easy to use a printed map, and this should be considered a fundamental security measure. Very often with this option you get the results that you want and you avoid the hassle of synchronising electronic devices. We already saw how to read coordinates on a map earlier.

2. Using loaded maps on the GPS

This option is useful to get a general picture of your whereabouts and not much more because the GPS screen is very small and the quantity of information it can show without saturation is very limited.

Some GPS devices come with maps already loaded from the manufacturer such as the maps from MapSource (Garmin), MapSend and others you can be purchased. These are very general road maps with some points that might interest you as the image shows.

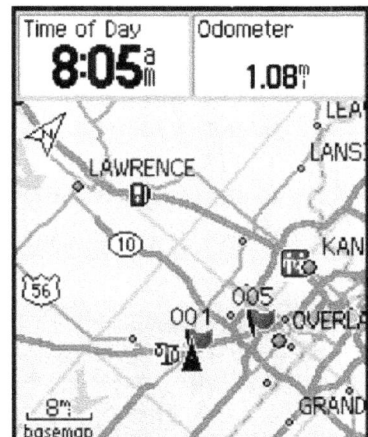

Another alternative is to create or obtain a map with the things that you need and upload it to the GPS. The map can be created with simple tools like GlobGPS or more complex ones like GPSMapper.

In my experience this option isn't very useful apart from very specific cases (the reason for it being mentioned) and it can be very problematic:

- Loading maps that aren't original and that don't have a standard protocol can mess up the GPS' software and damage it.

- When loading a map all the other maps on the GPS are deleted apart from the base map.

- More importantly, you run the risk that the tools take up all your attention: you could end up ignoring the more important things or become over-dependent on these maps, which would lead you to work with a bold unfamiliarity of the field

If you still think it's important, you will find more information on how to do it in reference 10 of the bibliography (Carlos Puch) and with an internet search.

3. Using maps loaded on a computer

Sometimes you have a map that is more interesting for your pursuits than the ones that Google Earth offers you. If you have it digiliatized or if you can digiliatize it maybe you can load it on Google Earth as an **image overlay**. The fundamental requirement to be able to use it is that it has a simple cylindric projection or WGS84 with the north upwards. If your map is like this, it will have the meridians and the parallels equidistant, parallel and crossing each other in right angles. For small areas, UTM maps can be used without any problems by ajusting some settings.

In the image a part of a 1:50,000 map has been loaded over the satellite image of the Teide to be able to see contour lines. Note how the map is represente with 3D relief:

Try not to load images that are too big so as not to saturate your computer memory.

If you don't have a map that fits these requirements, you can add it in other programs like Ozi Explorer or CompeGPS. In those cases you will need to **calibrate it**, in other words, the program will guide you through an easy process in which you will choose points in the image and introduce their coordinates. The program's manual will explain how to do so.

Overlaying maps

Add a general cartographic map to part of an Indonesian island between coordinates 5° 36' and 5° 38' north and 95° 7' and 95° 9' east.

1. Download the map here:

    ?

    www.arnalich.com/dwnl/goopsen/indonesia.jpg

2. Verify if the map meets the requirements: it has the north orientated upwards and the meridians and parallels are equidistant, parallel and orthogonal.

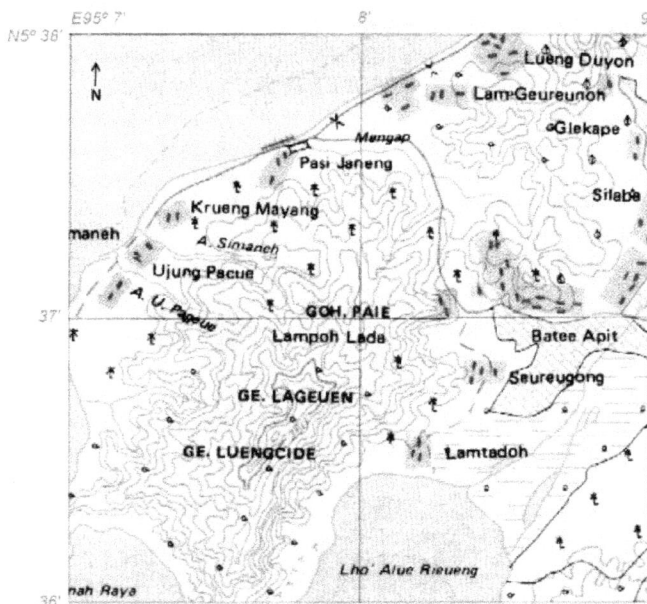

3. Google Earth will ask you for the coordinates of the corners of the image. For this it is important that you crop the image so as to have just a map from corner to corner, without borders or blank spaces. If you get stuck at this stage, you can download the already prepared image here:

    www.arnalich.com/dwnl/goopsen/indonesiax.jpg

4. The coordinates of the map come in degrees, minutes and decimals of a minute format. To enter coordinates using this format, go to > Tools / Options. In the 3D View tab select Degrees, Decimal Minutes and click Apply:

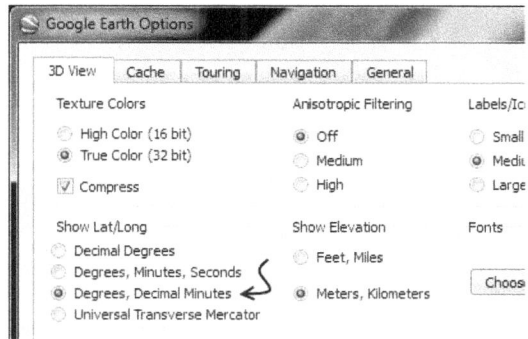

5. In Google Earth go to > Add / Image Overlay. Click the button Browse, and navigate to the location of the cut out image. This will make the image appear superimposed on the map within a green grid:

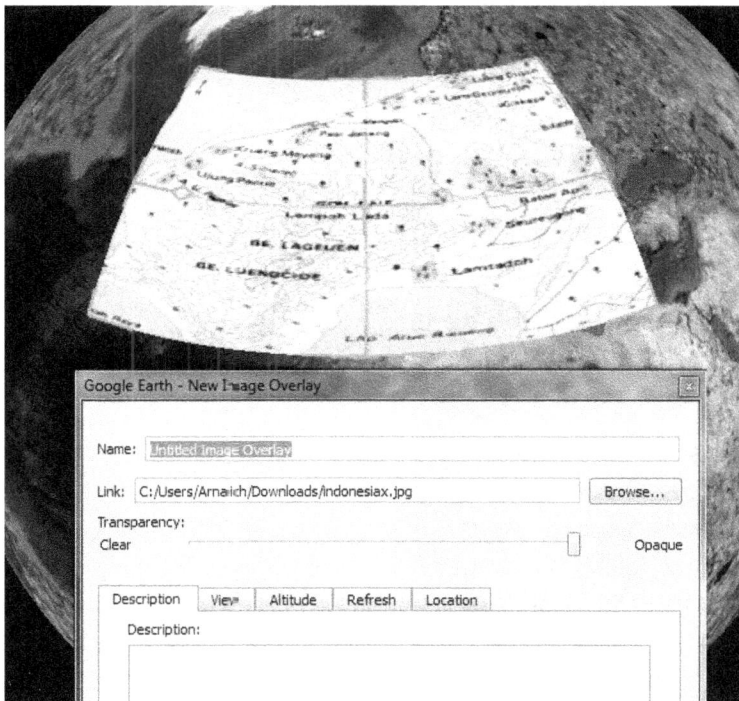

6. Go to the tab Location and introduce the coordinates of the extremes of the image that you have cut out. To avoid getting dizzy with automatic changes, begin by introducing the north and east values and then after, the rest. Accept to close the dialogue box.

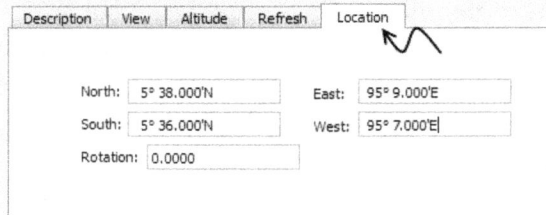

Description	View	Altitude	Refresh	Location

North:	5° 38.000'N	East:	95° 9.000'E
South:	5° 36.000'N	West:	95° 7.000'E
Rotation:	0.0000		

7. Chances are that you won't see anything yet. To zoom in to the area of the image that you just added, double-click on Untitled Image Overlay in the Places box.

Note that the alignment isn't very good, the coastline on the map and the satellite image don't coincide. This must be because the projection is UTM and not flat cylindrical.

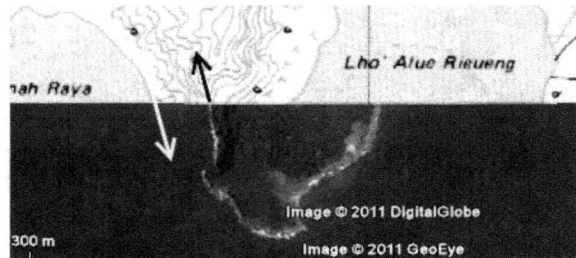

8. To solve this, right-click over the Untitled Image Overlay in the Places box and choose Properties. The dialogue box opens again and a grid with green crosses appears on the image overlay again. Click on one of them, and without letting go, drag the different parts of the map one after another until you have the best possible fit.

The degree to which you manage to adjust the image depends on the projection that you have, on the area that you cover, on small topographical errors, changes in the coastline, rivers, etc. that occur over time. Some maps adjust very well while others, like this one, with more difficulty.

3

Create

In this section we are going to cover the basic steps needed to create a map similar to the Haiti wells map that you saw in the first chapter. To create the map we've decided to use GPSvisualizer (even though this choice is debatable) over the other option, Google Earth Outreach: http://earth.google.com/outreach

Mapping the conditions of wells after a Tsunami

You're in the city of Meulaboh (Indonesia) after the 2004 tsunami. The tsunami has destroyed a big part of the coastal area. One of the main problems is that a large number of the drinking water wells aren't up and running. Some have been destroyed, others damaged and others covered in salt water and rubble that make them unusable until they're cleaned.

Downloading and converting GPS data

6

Starting from a GPX file, create another CSV file that GPSvisualizer can use.

1. Download and install the free GPSbabel program if you didn't do so earlier:

   ? www.gpsbabel.org/download.html

2. Download the GPX file that contains the data collected by the field survey team. You need it as a file, if it opens in your web browser instead, try right clicking and selecting Save as:

   ? www.arnalich.com/dwnl/goopsen/meulaboh.gpx

3. Open GPSbabel and:

   3.1 Look for the meulaboh.GPX file wherever you saved it by entering the name of the file in the input part of the dialogue box.

   3.2 Select the type of file as GPX XML for input.

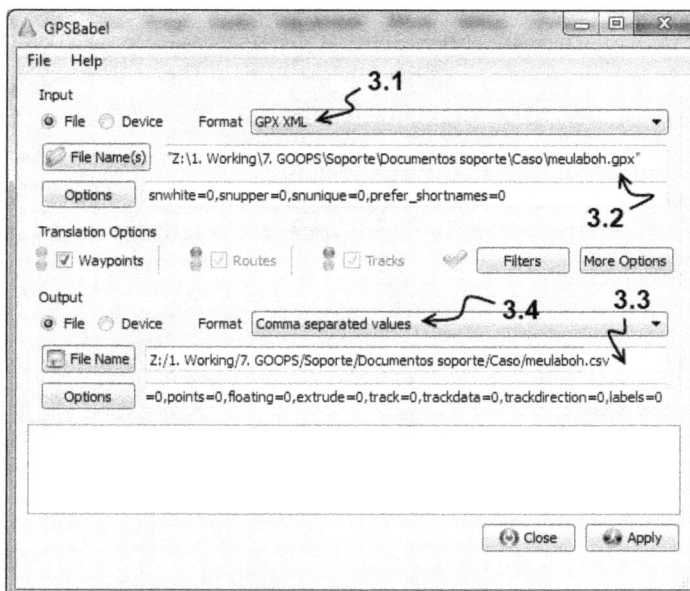

3.3 Click on the name of the output file and browse until you find the folder where you want to save it. Then write a name, for example Meulaboh and add .csv at the end (in other words meulaboh.csv) so that the system will recognize it.

3.4 Select the type of output file as Comma Separated Values.

When you're done press Apply to end the conversion. The file should appear in the folder that you decided to save it in.

4. Open the file with Excel or a similar program. To be able to see the file, you will probably have to select All files in the drop down menu of Open:

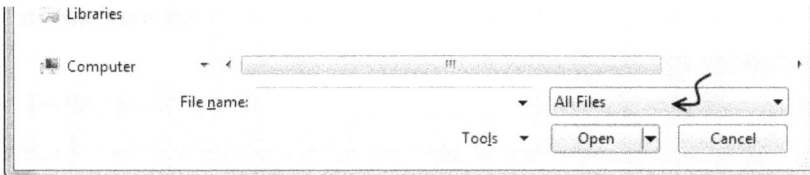

It will look similar to this:

The next step is to separate the data into columns, as initially they are all separated by commas within column A.

5. Select the whole column A by clicking on the A in the heading. Once done, go to the Data tab and select Text to Columns:

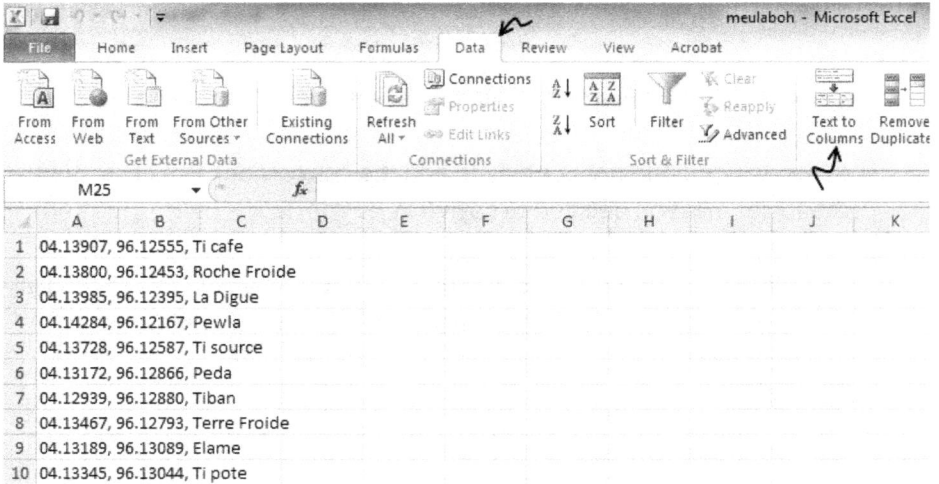

6. Select Delimited in the first step of the assistant:

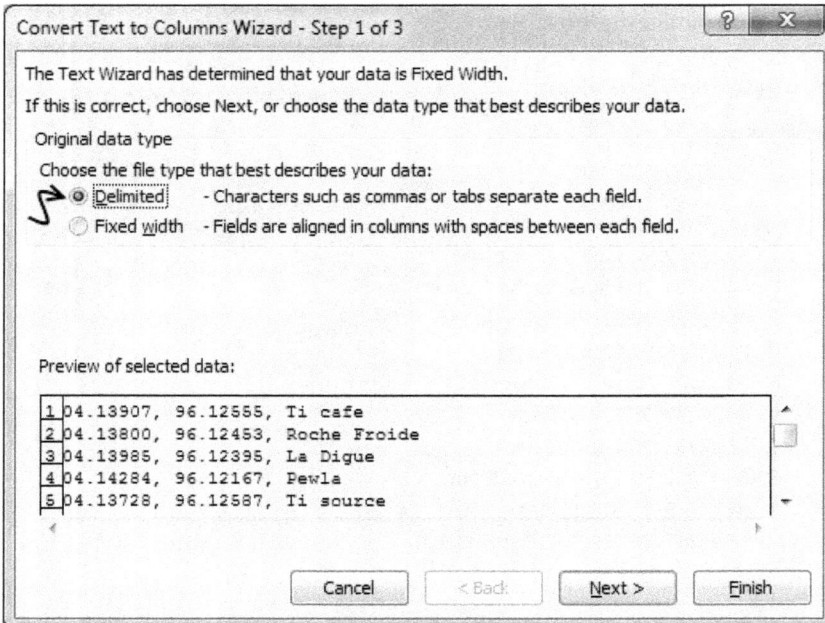

7. In step two check Comma and uncheck the rest of the options:

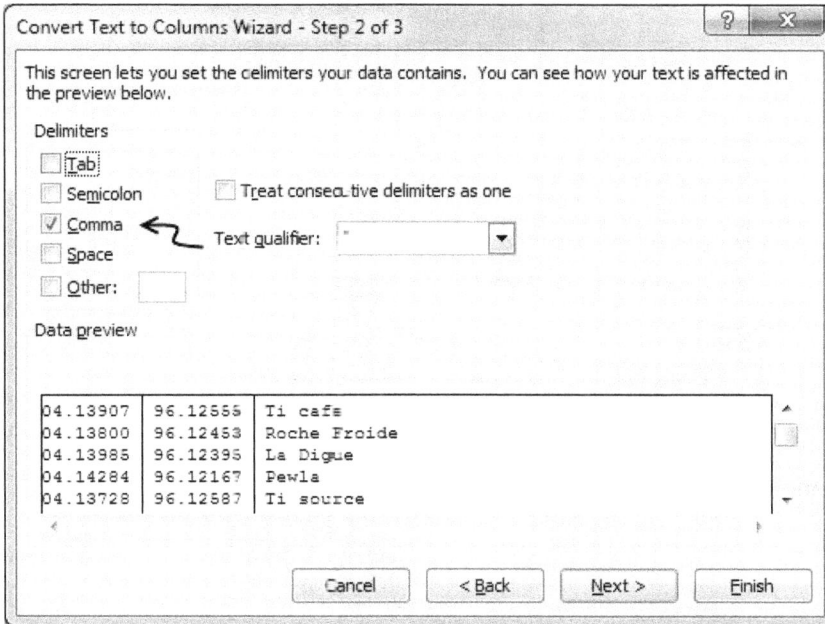

8. In step three, press Advanced...:

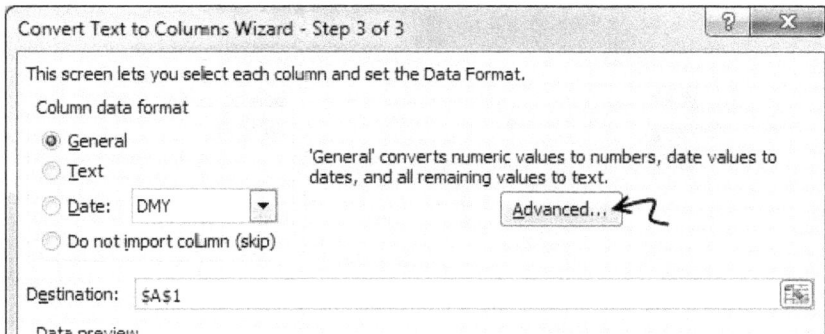

Pay close attention to the next steps if you don't want to go crazy figuring out why your file isn't recognized or waste time recovering points in the middle of the Pacific Ocean and other unlikely places. **The decimal separator should be a dot.**

9. Select . as the decimal separator and ' as the thousands separator and accept:

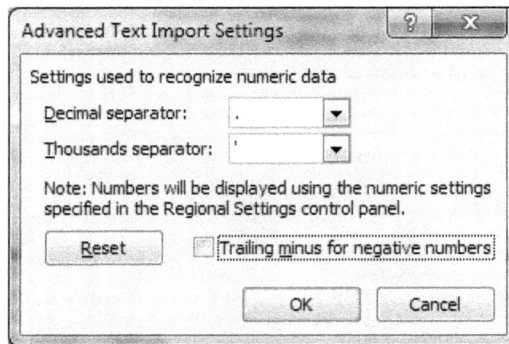

10. In the Excel file menu, select Options:

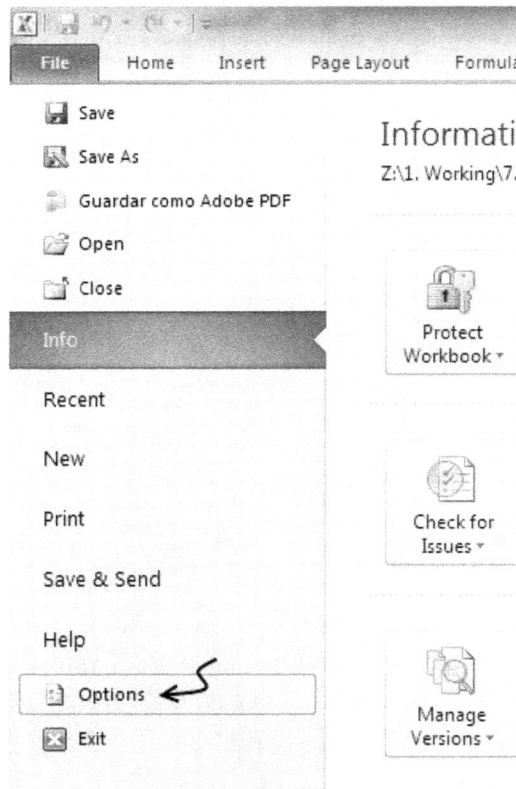

11. In Advanced, make sure you have the same selection:

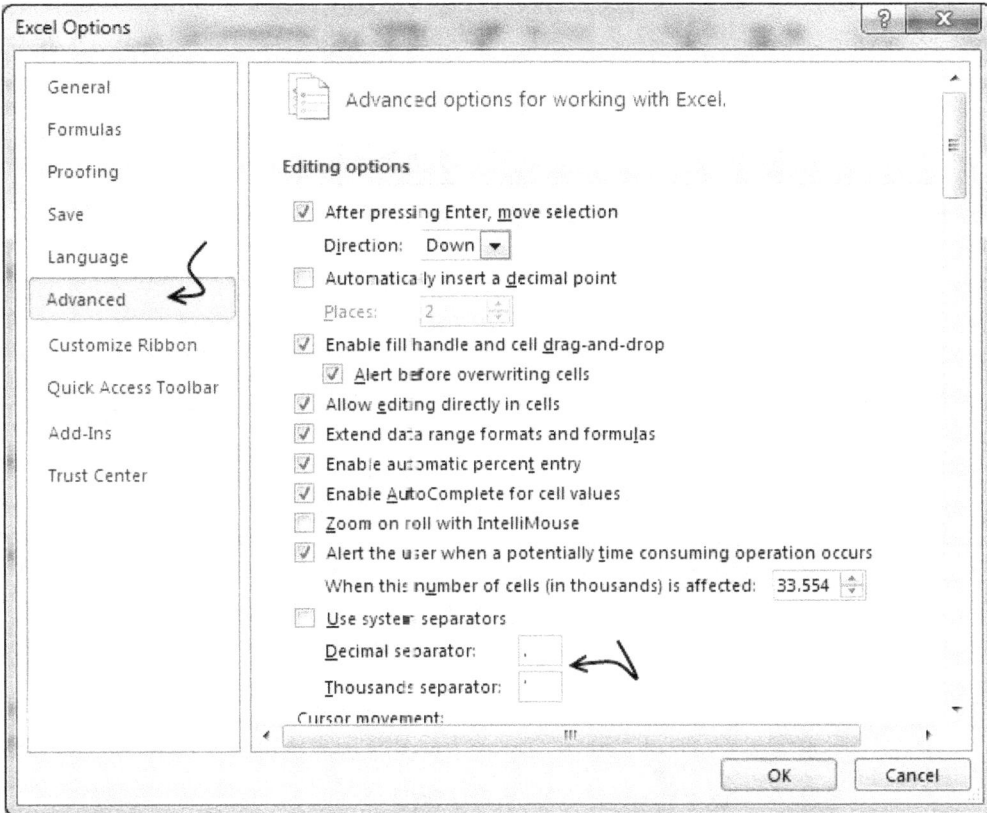

The steps in different spreadsheets and operative systems differ and it is impossible to cover them all here, but you now have an idea of what you have to look for.

12. Add a new row on top of your table so that GPSvisualizer knows what each column is. Type latitude, longitude and name in the heading cells, exactly that: "latitude", "longitude" and "name".

13. Save the file in CSV format (values separated by commas).

14. Navigate to www.gpsvisualizer.com and select Google Earth KML:

15. Press Browse, select the file created in step 13 and press Create KML file:

If everything went well, a link will appear to download or directly open the file in Google Earth:

GPSVisualizer

Google Earth output

Your GPS data has been processed. Here's your KML or KMZ file:

1335904388-02289-41.203.179.34.kmz

If you've already installed Google Earth, clicking the above link should oper problem.

The result is this:

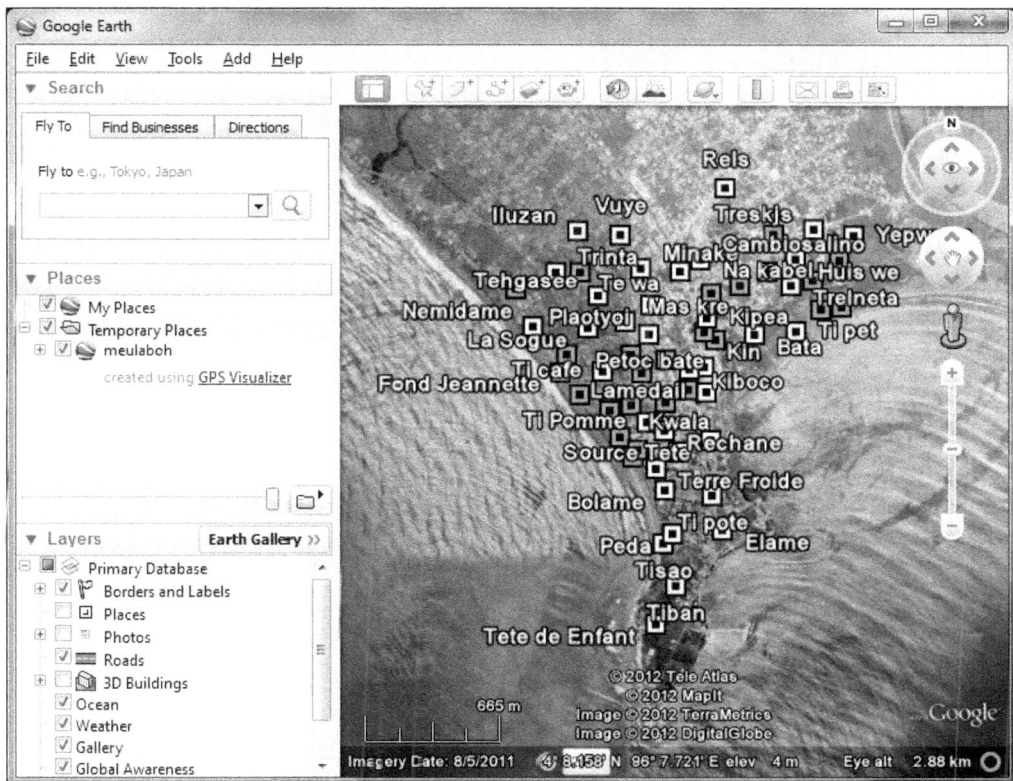

If something went wrong, run through this short checklist of most common errors:

1. Maybe your GPS is saving coordinates in a format other than decimal degrees (i.e. 46.95848°).
2. If you have points in a very distant location, you probably have some points where the thousands separator has been interpreted as a comma.

3. If GPSvisualizer gives you an error "No valid GPS Data detected!" and the previous explanations are of no help, check that the columns of your index are headed with "name", "longitude" and "latitude" without the quotes and that there are no typos.

GPSVisualizer

No valid GPS data detected!

If you think you really did upload a valid GPS file, please let me know and I'll look into it.

Often, this error comes up because you didn't include a header row at the beginning of your tab-delimite which! For example, GPS Visualizer has no idea what this means:

```
45.40252,-121.77589,1399,point 1
45.39254,-121.69946,2394,point 2
```

Whereas the following makes perfect sense, because there's a header row that defines the fields:

```
latitude,longitude,elevation,name
45.40252,-121.77589,1399,point 1
45.39254,-121.69946,2394,point 2
```

If for any other reason you are stuck and you don't understand why, you can download the final file for this exercise here:

www.arnalich.com/dwnl/goopsen/ex6.zip

Adding color to the position markers

7

Once the columns have been added, show the markers according to the condition of the well (destroyed, to rehabilitate or functioning).

You can add as many columns as you need between the column name and those of longitude and latitude. You can add color to your points by creating a column with the title "color" (Not colour!). The color cell for each row should contain the name of a color in English (red, blue, green, white, black, yellow, pink...) depending on the variable that you want to represent.

To make this change automatically in Excel you can add a conditional function similar to this one:

=IF(C2="ko";"red";IF(C2="rehab";"yellow";IF(C2="ok";"green")))

In other words, if the well is KO it will be red, if it needs rehabilitation it will be yellow and if it is OK it will be green.

1. Download the file with the columns ready:

www.arnalich.com/dwnl/goopsen/meulaboh7.csv

2. Add a column named color:

	A	B	C	D	E	F
1	name	color	condition	latitude	longitude	
2	Ti cafe		ok	4.13907	96.12555	
3	Roche Froide		rehab	4.138	96.12453	
4	La Digue		ko	4.13985	96.12395	
5	Pewla		ok	4.14284	96.12167	

3. In the first cell, type in the conditional function and press Enter:

=IF(C2="ko";"red";IF(C2="rehab";"yellow";IF(C2="ok";"green")))

	A	B	C	D	E	F	G
1	name	color	condition	latitude	longitude		
2	Ti cafe	green	ok	4.13907	96.12555		
3	Roche Froide		rehab	4.138	96.12453		
4	La Digue		ko	4.13985	96.12395		
5	Pewla		ok	4.14284	96.12167		
6	Ti source		rehab	4.13728	96.12587		
7	Peda		rehab	4.13172	96.12866		
8	Tiban		rehab	4.12939	96.1288		
9	Terre Froide		ko	4.13467	96.12793		

Since the value in the cell C2 is "ok" it will take the value "green". If the condition column had been in the column E, for example, the formula would have used E2 instead of C2.

4. Drag the formula to the rest of the column. To do so, put the cursor on the lower right corner until it turns into a black cross, then click and, without letting go, drag until the last line. The cells automatically take the values:

	A	B	C	D	E	F
1	name	color	condition	latitude	longitude	
2	Ti cafe	green	ok	4.13907	96.12555	
3	Roche Froide	yellow	rehab	4.138	96.12453	
4	La Digue	red	ko	4.13985	96.12395	
5	Pewla	green	ok	4.14284	96.12167	
6	Ti source	yellow	rehab	4.13728	96.12587	
7	Peda	yellow	rehab	4.13172	96.12866	
8	Tiban	yellow	rehab	4.12939	96.1288	
9	Terre Froide	red	ko	4.13467	96.12793	

5. Save the file (as CSV!).

6. Go to GPSvisualizer and select Google Earth KML like you did in the last exercise.

In the next steps you are going to improve the look of the map by increasing the precision and avoiding the saturation in the image:

7. In the next screen, in Waypoint options, select Paddle as symbol, which is more precise than the default square icon:

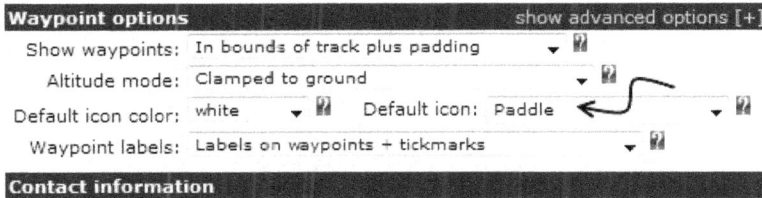

8. In the dropdown menu immediately below, change the Waypoint labels box to Mouse-overs only. This gets rid of the permanent labels that crowd maps.

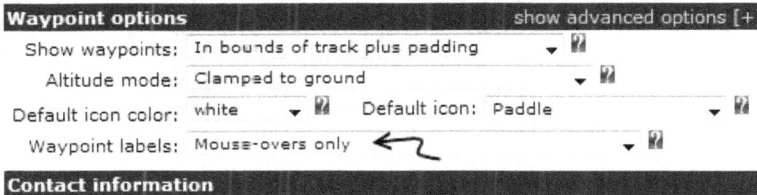

9. In Browse load the file that you have created in step 5 and press Create KML file.

Whilst the file source name is the same you don't need to load it again on Browse. The file and its location are remembered between uses. You just need to go back in your web browser and press Create KML file each time you have modified the source file.

Observe that once you save the csv files, they don't have the formula anymore, only the values that were generated.

The result is shown below. You can also download the results of this exercise here:

www.arnalich.com/dwnl/goopsen/ex7.zip

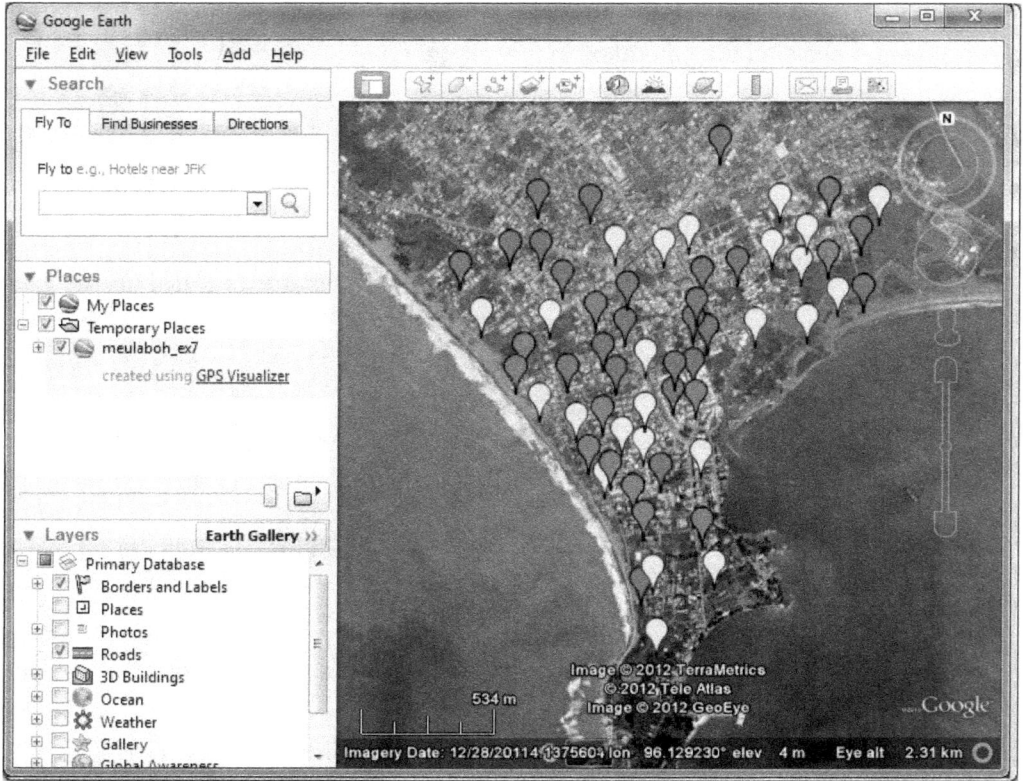

These are some of the color names you can use. If you have the printed black and white version of this book, you can obtain the image in:

www.arnalich.com/dwnl/goopsen/GEcolors.png

aqua	navy	lime
black	olive	magenta
blue	orange	maroon
brown	pink	white
cyan	purple	teal
fuchsia	red	violet
gray	silver	yellow
green	tan	

Adding a legend

Add a legend to the map from the previous exercise.

The legends are added in the form of a superimposed image. First you will have to create an image with the legend, the title, and any other information that is necessary and then link it with a KML file that Google Earth recognizes.

1. Prepare the image that you need on any image editing program of your choice. In this exercise, we are going to use the image shown on the right.

   Don't try to do very big legends because they will cover too much of the screen.

   Also don't take too much care, because unfortunately Google Earth makes them smudgy unless you use the superoverlay function which is laborious and complex.

   Well condition after Tsunami in Meulaboh

   01/13/05

   In working order

   Rehabilitation needed

   Destroyed

   Fictional data without borders

   All data used in these exercises is fictive

2. The second step is to upload it to an internet server so that it is available for everybody who opens the file. You can try a number of services for this, Flickr, Picassa, etc. In this case, the image is already in a server with the following route:

   http://www.arnalich.com/dwnl/goopsen/legend.png

3. Open a blank document in Windows Notepad; you can find it in > Start / Accessories / Notepad:

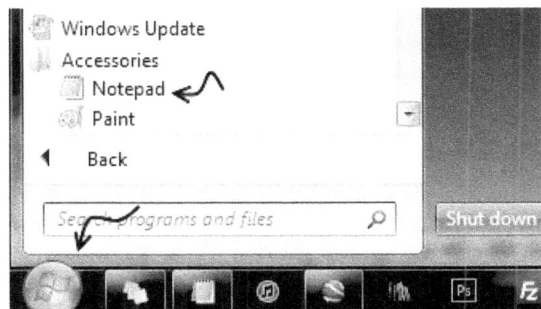

   Windows Update
   Accessories
   Notepad
   Paint
   Back

   Search programs and files
   Shut down

4. Paste the following code, that can be downloaded at:

www.arnalich.com/dwnl/goopsen/legend.txt

```
<?xml version="1.0" encoding="UTF-8"?>
<kml xmlns="http://www.opengis.net/kml/2.2">
  <ScreenOverlay>
    <name>Legend</name>
    <Icon>
      <href>http://www.arnalich.com/dwnl/goopsen/legend.png</href>
    </Icon>
    <overlayXY x="0" y="1" xunits="fraction" yunits="fraction"/>
    <screenXY x="0" y="1" xunits="fraction" yunits="fraction"/>
    <rotationXY x="0" y="0" xunits="fraction" yunits="fraction"/>
    <size x="0" y="0" xunits="fraction" yunits="fraction"/>
  </ScreenOverlay>
</kml>
```

Note that in <href> you must enter the file path for the image. In a real case you would put the URL for where you uploaded the image:

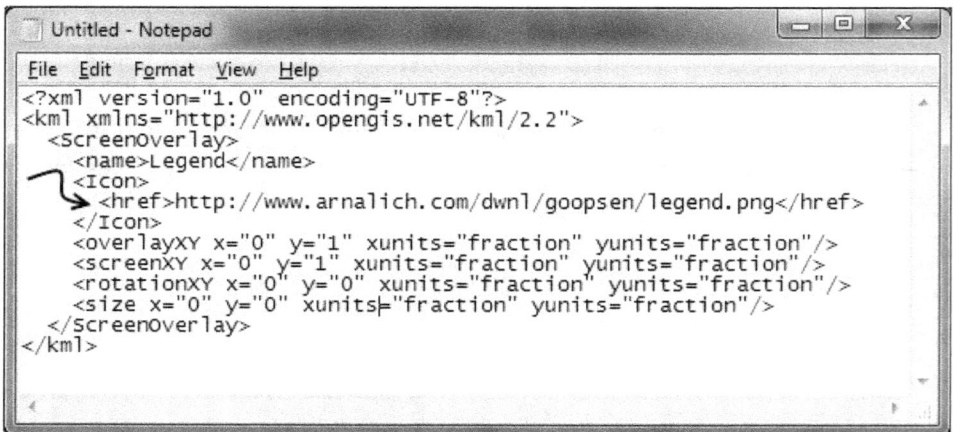

5. Save the file as KML. To save with this file type, in the Save as type dropdown menu choose All files option and add .kml to the file name:

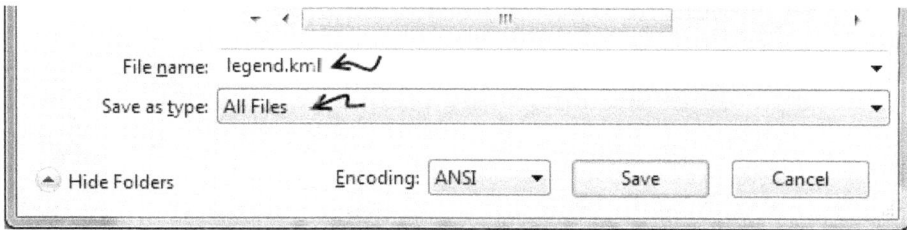

6. Open the file created in Exercise 7 with Google Earth and also open the legend file. Be patient because the image can take some time to download from the server. This is the result:

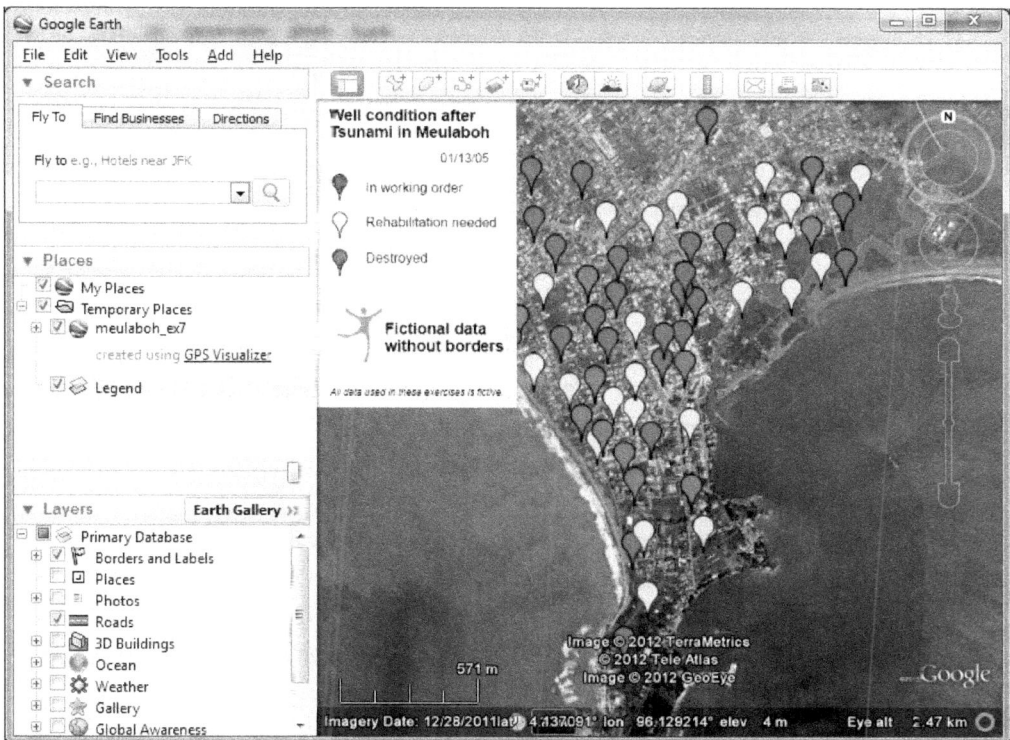

7. It can be useful to unite the legend and the map in one file. To do this, click on Temporary Places to select it and then go to > File/ Save/ Save Place As:

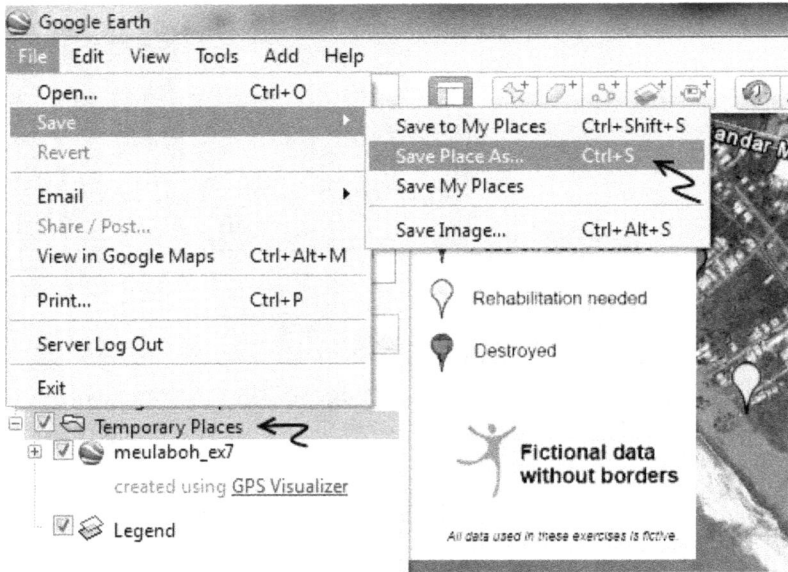

Note that the image quality is reduced a little bit as part of the process. Adding transparency to the image, like the legend used in exercise 1, may be a good idea.

Make sure that the legend includes all the necessary information to make it useful. A map with colored markers and nothing else is completely useless, as nobody knows what is what, if the information is still relevant, etc. As a quick checklist, check if your legend:

- ✓ Has a title that communicates a clear idea.
- ✓ Identifies the organization that produced it.
- ✓ Is dated and indicates if it is part of a series.
- ✓ Has an appropriate disclaimer, for example, regarding recognition of frontiers.
- ✓ Mentions the sources of the information used.
- ✓ Explains the map.

You can download the results of this exercise here:

www.arnalich.com/dwnl/goopsen/ex8.zip

Adding color according to quantities

Create a map of well salinity to know which ones need to be bailed out with a pump to remove the sea water.

The case is very similar to that of Exercise 7 only that this time we are going to have colors depend on numerical values. For the salinity we are going to consider that values under 750 were normal before the tsunami, from 750 to 1500 are higher but the water is still drinkable and above 1500 the water saline.

To make this change a conditional function similar to this one is added in Excel:

=IF(C2<750;"aqua";IF(C2<1500;"teal ";IF(C2>=1500;"black";"")))

In this function the conditions are nested one within the other. It works in the following way: first it asks if C2 is less than 750. If it's true, it notes the color aqua, if it's false, instead of putting another value, it asks again and like this you can successively incorporate as many intervals as you want. The tail;"" at the end is so that it's left blank in case of there being an error.

1. Download the file with the information that you need:

    ?

    www.arnalich.com/dwnl/goopsen/meulaboh9.csv

2. Add a color column.

3. Type the formula into the first cell and extend downwards.

    =IF(C2<750;"aqua";IF(C2<1500;"teal ";IF(C2>=1500;"black";"")))

4. Save the file so it can be used with GPSvisalizer.

5. Create the KML with GPSvisalizer.

6. Create the legend. The URL to the image is:

   www.arnalich.com/dwnl/goopsen/salinitylegend.png

7. Unite the map and legend in one file. Try and give it the name "Well salinity" in the Places area:

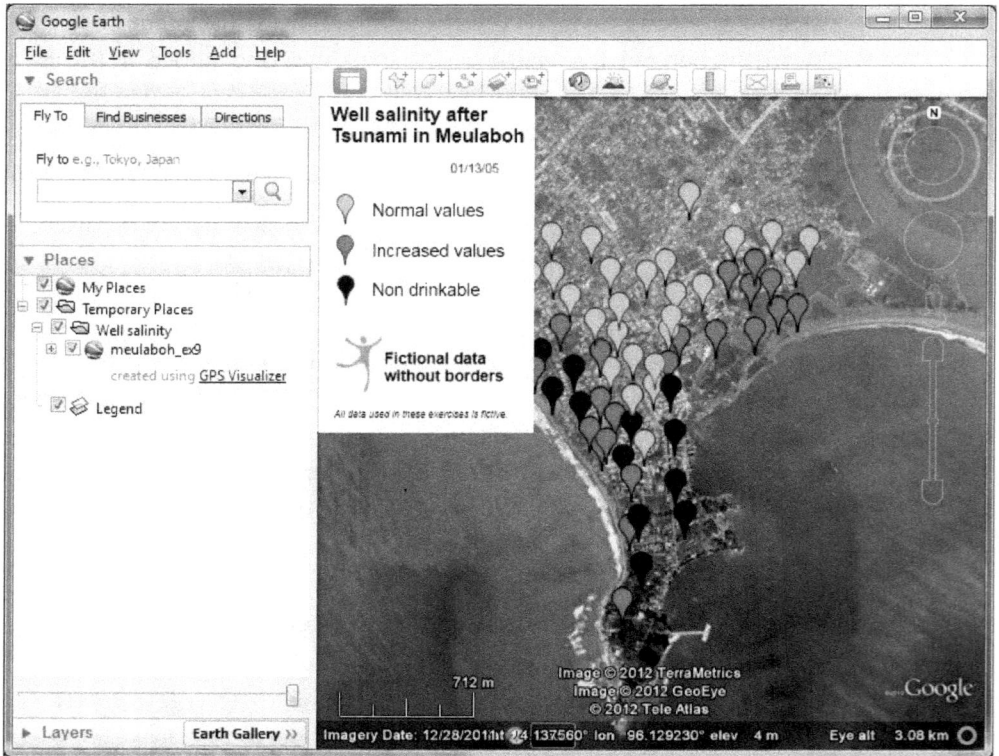

With a map like this it's a lot easier to plan. For example, you could propose to first clean the wells displayed in black leaving a certain distance between them so that the population has at least one well nearby that has drinkable water.

You can download the results of this exercise here:

   www.arnalich.com/dwnl/goopsen/ex9.zip

Classifying in folders

Create a map that classifies wells according to their turbidity and shows their condition using colors.

To create folders that group together points with a certain condition we have to create a new column with the title "folder". The water is classified as clear if it has less than 5 NTU and therefore can be chlorinated efficiently, and as turbid if it has more.

1. Download the file with the information that you need:

www.arnalich.com/dwnl/goopsen/meulaboh10.csv

2. Create a new column and type folder in the first cell:

	A	B	C	D	E	F	G
1	name	color	folder	condition	turbidity	latitude	longitude
2	Ti cafe	green		ok	45	413.907	9.612.555
3	Roche Frc	yellow		rehab	65	4.138	9.612.453
4	La Digue	red		ko	200	413.985	9.612.395
5	Pewla	green		ok	300	414.284	9.612.167
6	Ti source	yellow		rehab	120	413.728	9.612.587
7	Peda	yellow		rehab	50	413.172	9.612.866

3. Create and extend the formula that assigns the turbidity values in the folder column.

=IF(E2<=5; "clear"; "turbid")

4. Create the KML file with GPSvisualizer.

5. Open the file and press + near the file you just loaded in the Places box:

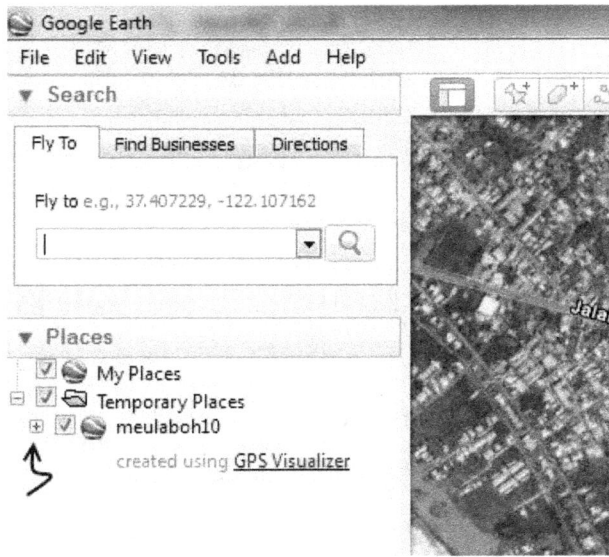

By pressing the + the two folders that you have created are shown:

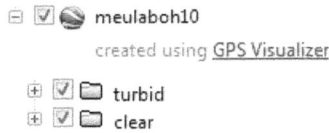

If you uncheck turbid box, all the murky wells will disappear from the map.

10. Change the name of the layer to "Condition vs. Turbidity".

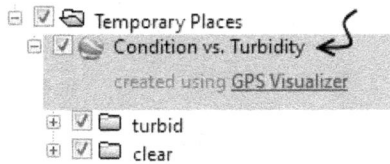

11. Add the legend for the condition of the wells and save the resulting file. The result is shown on the next page and you can download it at:

www.arnalich.com/dwnl/goopsen/ex10.zip

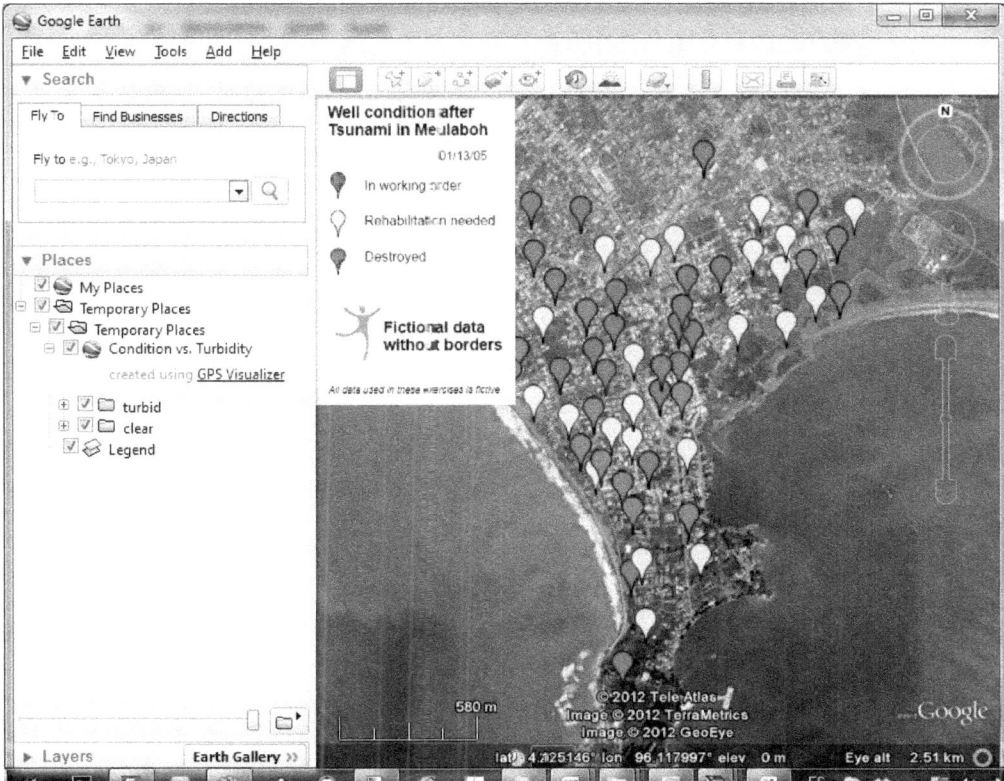

Adding sub-folders

Create a map that classifies the wells depending on both their turbidity and salinity and shows the condition with colors.

To create folders inside of folders add a backward slash (\) between the values inside the cell. For example, the content of cell C32 is turbid\fresh.

To make this change automatically a linked function, similar to this one, is added in Excel:

=CONCATENATE(IF(F2<=5; "clear"; "turbid");"\";IF(E2<=1500; "fresh"; "brackish"))

This formula looks for the corresponding turbidity value, adds \ and then looks for the corresponding salinity value. To simplify things, we are only interested in knowing if the water is sweet or too salty to drink.

1. Download the file with the necessary information here:

    www.arnalich.com/dwnl/goopsen/meulaboh11.csv

2. Add a folder column.

3. Type the formula in the first cell and extend it downwards.

    =CONCATENATE(IF(F2<=5; "clear"; "turbid");"\";IF(E2<=1500; "fresh"; "brackish"))

If everything went right the values will appear separated by slashes:

	A	B	C	D	E	F
	name	color	folder	condition	salinity	turbidity	lati
1							
2	Ti cafe	green	turbid\brackish	ok	4350	45	4.
3	Roche Frc	yellow	turbid\brackish	rehab	4750	65	
4	La Digua	red	turbid\brackish	ko	4550	200	4

C26 =CONCATENATE(IF(F26<5; "clear"; "tu

4. Create the KML file.

5. Open the KML file and add the legend.

6. Save it all as one single file. Press + to open the subfolders:

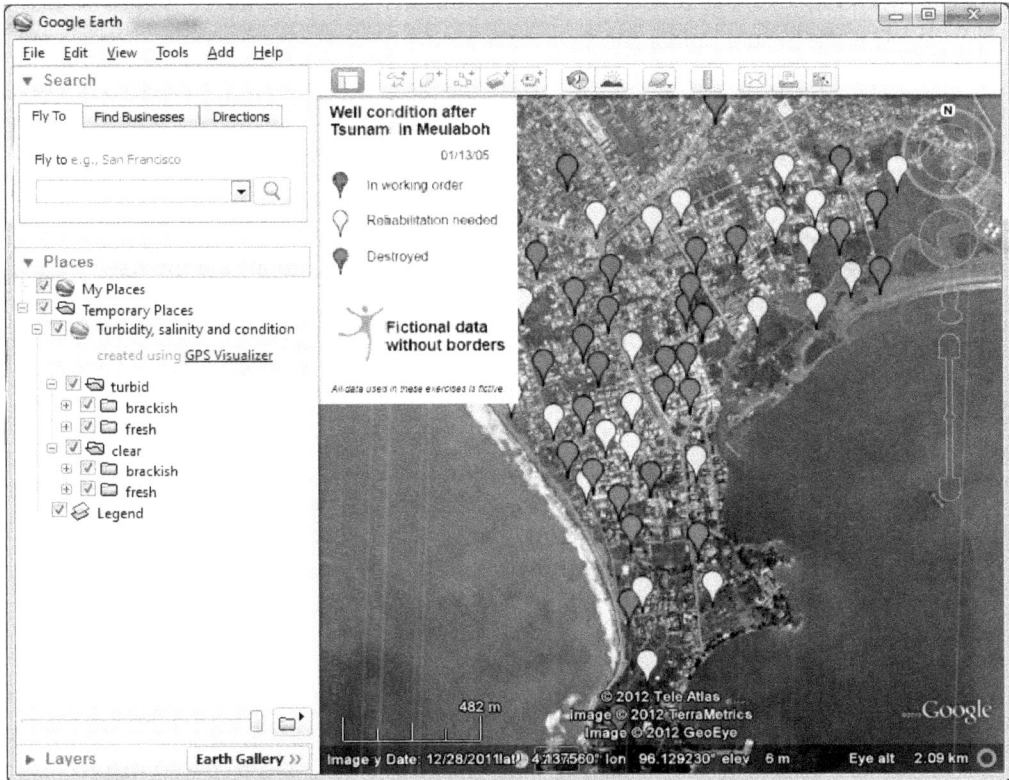

You can download the resulting files at:

www.arnalich.com/dwnl/goopsen/ex11.zip

Creating maps for a website

Create a web map that shows the size of the refugee camps.

Sometimes you want to make sure that anyone, with Google Earth installed or without, can see a map online and interact with it. In those cases creating a KML file would not be the best option.

1. Download the file with the columns already prepared:

   www.arnalich.com/dwnl/goopsen/meulaboh_camps.csv

2. On www.gpsvisualizer.com this time click on Google Maps:

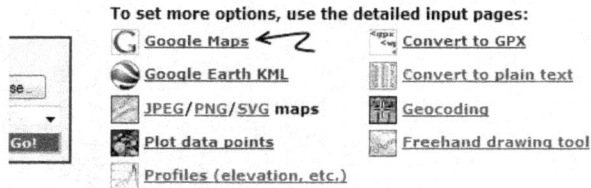

3. Select Quantitative data:

4. Press Browse and select the source file that you have just downloaded.

5. Fill out the boxes that are shown by arrows as in the image so that the size of the point and the color depend on the population of each camp.

Data point options   show advanced options [+]

Colorize using this field: custom field   ▼ Min: Max:

Custom colorization field: Size

Colorization legend: Bottom left ▼   Legend steps: 5

Lightness: 80% ▼   Saturation: 100% ▼

Spectrum direction: up ▼ Hue 1: 0° ▼ Hue 2: 240° ▼

Color of values beyond min. or max.: Gray   ▼

Resize using this field: custom field   ▼

Custom resizing field: Size

Minimum radius: 6   pixels   Maximum radius: 16   pixels

Unresized radius: 8   pixels   Single-point map width: 10 km

Calculate frequency: No   ▼ (creates a field called "N")

Show point names: Yes ▼   Show point descriptions: Yes ▼

Default marker color: red   ▼

6. Press Draw the map to see the result:

Google Maps output

Your GPS data has been processed. Your Google Map should be displayed below, and it's also **temporarily** available to view or download f
please contact me and explain the problem. If you want to save your Google Map to your Web site, the HTML source of the map must be m

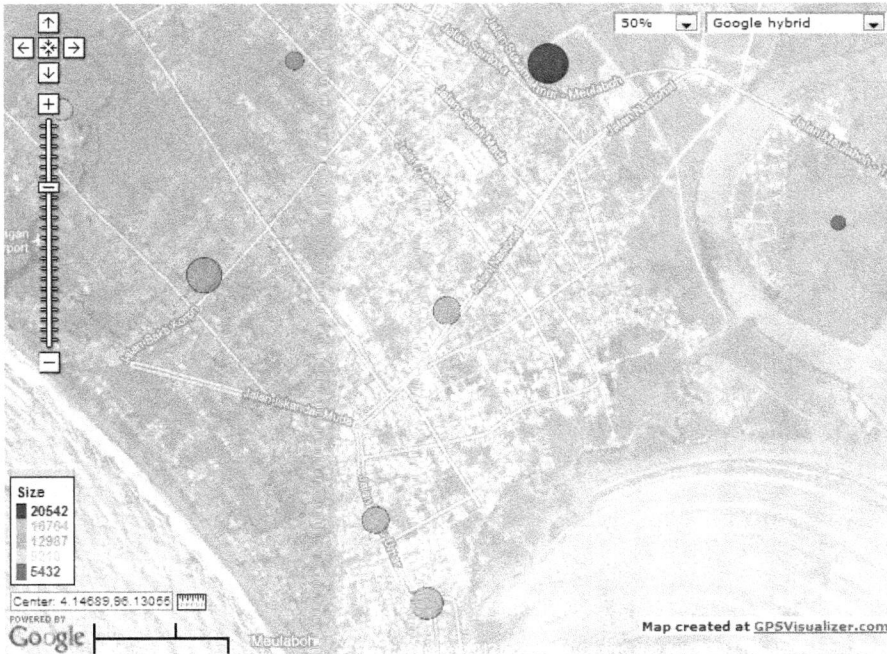

By clicking on download you can download the map to incorporate it in any webpage.
You will need a Google Maps API key to keep it working after a while. The internet or
even GPSvisualizer will explain you how to do that.

Displaying data within the info balloons

13

Add information to the point's balloons of the map created in exercise 11.

When you click on one of the markers an informative globe appears that shows the information available about that point:

As you can see the balloon is empty right now, it is not very informative. In this exercise, we will learn how to show all the data about the point the balloon: turbidity, salinity, etc.

1. On GPSvisualizer, in Browse load the file meulaboh11.csv that you downloaded in exercise 11.

2. In Waypoint options, press + to unfold the advanced options menu:

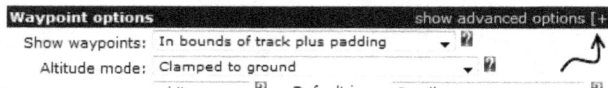

In the Synthesize description field goes a small code that you will prepare in a moment that will add the data to the balloons.

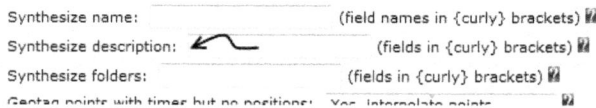

3. Open any text processor, for example, Word.

You are going to repeat the following code for each column that you want to show:

> <b>Field Name</b> {column name} Units

The labels <b> and </b> put in bold what is between them and <br> inserts a line jump.

For the first column the field name is Condition and the column in Excel is called by the same name. There aren't any units. The resulting code is:

> <b>Condition: </b> {Condition}

The second column is salinity. Once added to the previous code, the updated code would become:

> <b>Condition: </b> {Condition}
 <b>Salinity: </b> {salinity} microS/m

The last one is Turbidity, with units in NTU:

<b>Condition: </b> {condition}
 <b>Salinity: </b> {salinity} microS/m
<b> Turbidity: </b> {turbidity} NTU

4. Copy this code to the Synthesize description field and create the KML file. Click on one of the points to see its properties:

Download the results on: www.arnalich.com/dwnl/goopsen/ex13.zip

4

Share

Sharing the maps you created is very simple. There are numerous ways and each organization or person will prefer one over another. The idea is not to make an exhaustive list of possibilities because those would change within a few months as the ways of working and technologies evolve. Rather, it is to present some of them so that you can investigate the one that is most convenient for you.

Before sharing

Don't share:

- Data that isn't yours or that you don't have permission to share.
- Sensitive data.
- Information that could put people in danger.
- Information that puts people's privacy at risk.
- …

Apart from these common sense recommendations, remember that a map is a document for the communication of ideas. The more logical and understandable your map is, the more useful it will be. Avoid some frequent errors:

1. **Over-complicated maps.** A map is used for communicating _one_ idea. That idea is in the title. It includes what is needed and avoids a diarrhea of unnecessary ideas and information that misinforms.

2. **Incognito maps.** Maps without fundamental data like dates, agencies, legends, etc.

3. **Camouflage maps,** where you can't see anything due to errors in the scale, the text size or colors use (you can see everything beautifully on the computer screen but not in the reports that are printed in black and white.

4. **Fickle maps.** If you are mapping the changes of a region over time avoid each map of the series having different colors, region span, symbols, etc. The series must have the same legends to be able to quickly interpret the changes between each one.

5. **Maps of your misunderstanding.** In line with point 1, make sure that in addition to knowing and having a clear idea you want to get across, that you also do understand it.

A mailing list or a group

Sharing maps with a mailing list or group is probably one of the easiest options. The created files are sent, with the advantage that the user receives a notification when there are new files.

On the negative side:

- There isn't an efficient file and search system.
- With email inboxes looking every day more like a Tetris game it is very probable that they go unnoticed or are left for… never.
- It is only shared with certain users, not with the web.

Mailing lists have the potential to be profoundly irritating. If each time you send an email to 200 people you start to receive replies like "Hello, thanks for the file" or an avalanche of auto responders reminding you of everybody else enjoying their vacations, it will very quickly turn into a nuisance.

On the other hand, if you simply copy the email addresses you are exposing everyone on the list to share the inheritance of faraway princes, buy discount v1agra and all sorts of spam.

To avoid these inconveniences you can use some free services. Google, Yahoo!, and many others offer the possibility to create groups:

- https://groups.google.com/
- http://groups.yahoo.com/

When you choose a service don't make people waste their time with closed services such as Facebook (however universal you think they are). Try and use services that people are already familiar with and don't require new registrations or introductions.

If you don't want or need replies you can use some marketing services, the best one is probably www.mailchimp.com.

A blog

On the positive side:

- The contents are generally made public and accessible to searchers.
- They allow you to show the chronology of changing situations.
- They allow for contents preview.
- They easily accept a journalistic style that promotes comparing.
- Very practical for a unique theme, for example, following up on a disease.
- They can incorporate search functions.
- They allow you to create a feed that informs of new postings.
- They allow you to incorporate buttons to share social networks: twitter, Facebook, Google plus and more to spread the word.
- You can create and share interactive maps like this one from exercise 12: www.arnalich.com/dwnl/goopsen/ex12.html

On the negative side:

- If there are a lot of maps with different subjects or without a clear chronology it gets disorganized very quickly.
- The older postings are devalued.
- Improving the look requires basic knowledge of html.

There are many free blog services, the most common being:

- www.blogger.com
- www.wordpress.com

A website

The advantages are similar to those of the blog except for clear time lines. The main difference is the possibility of creating navigation systems and more complex files that don't end up being forgotten over time.

On the negative side:

- Improving the look requires both a basic knowledge of html and time.

An example is the U.S. Holocaust Memorial Museum web:
www.ushmm.org/maps/projects/darfur/

File hosting

There are lots of services that permit a folder on your computer to automatically synchronize with a copy on a server. This folder can be shared and synchronized within a group of people.

The good thing about this system is that it doesn't require any work, it is automatic. The bad thing is that you lose all the functions of sharing on the blog network or web page.

The most known option, besides Google Drive, is probably www.dropbox.com, which allows up to 2GB of free space. This site has lots of tutorials on how to use it and is very easy.

Basically, in a folder called Dropbox you can have another subfolder with the maps you want to share. For example, here the Meulaboh maps folder has been created and has a symbol with blue arrows which shows it's automatically synchronizing because you've added a map (or any other file):

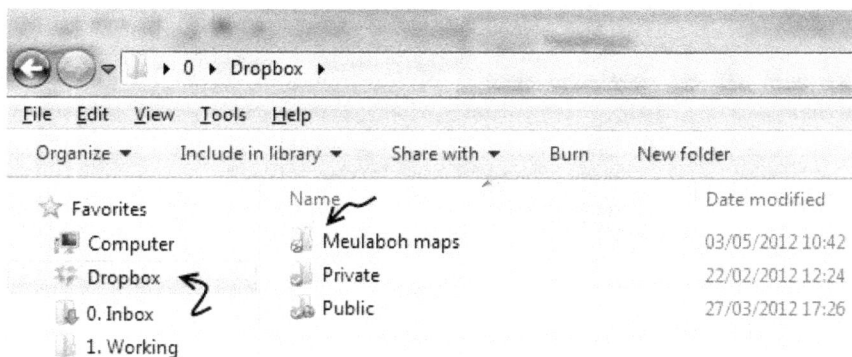

When you or the other users on the group open their computer the contents of the folder will be automatically updated.

You can also generate download links for the files and send the link to people rather than sending the file itself by email when it is too large.

5

Collaborate

Once again there are numerous forms of collaboration and each organization or person will prefer one over another. Here we are going to discover, using examples, how to do it with Google Docs without implying that it is better or worse. The procedures will be similar to those of other services.

Before collaborating

Collaborating requires people who are motivated and disciplined, work instructions, clear objectives and a series of rules to avoid chaos. Don't expect people to spontaneously collaborate in an organized manner. Some tips:

1. Agree on a **person responsible** or a general administrator.

2. Keep things as **simple** as possible. Be careful with the technophilia!

3. Agree on the **templates** of the data that is needed and in which order, to avoid having to manipulate the documents.

4. Collect **the least information possible**, only that which is genuinely useful and that can be collected **without gaps** and with a reasonable and sustainable effort even when the initial high motivation has faded.

5. Create a small document or **tutorial** that explains how to participate in a practical way and take time with each new collaborator to introduce them patiently to the system.

6. Create some **rules for naming documents and folders**. A system in which everything is in the folders called Maria or Maps, and the maps are called Test5.kml, RWS.kml is completely useless. The users should know what the map is about before opening it. An example could be:

    20070328 Salinity vs. Turbidity wells Meulaboh.kmz

    The first number is the date, which when expressed like this has the advantage that the maps are automatically organized by date in each folder.

7. Create a **Sandbox** folder, where the users can do tests without the risk of either breaking anything or that the results get confused with real data. Not many things can kill a system quicker!

8. Establish a routine of **security copies** adequate for the frequency of changes so as to avoid an involuntary swipe that ruins your efforts.

Collaborating with Google Drive

Google Drive (previously Google Docs) allows you to upload an Excel spreadsheet which numerous people, up to 50 at a time, can edit simultaneously and see the changes entered by other people in real time. The procedure is:

1. You need a Gmail account. If you already have one you have to access it, if you haven't you can create it on http://mail.google.com.

2. Select Documents on the upper bar:

   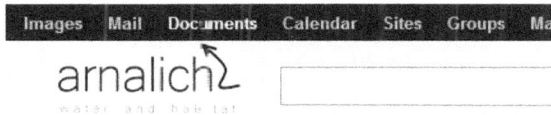

3. Click on the icon to upload the file and select Files. Afterwards navigate to the data file.

   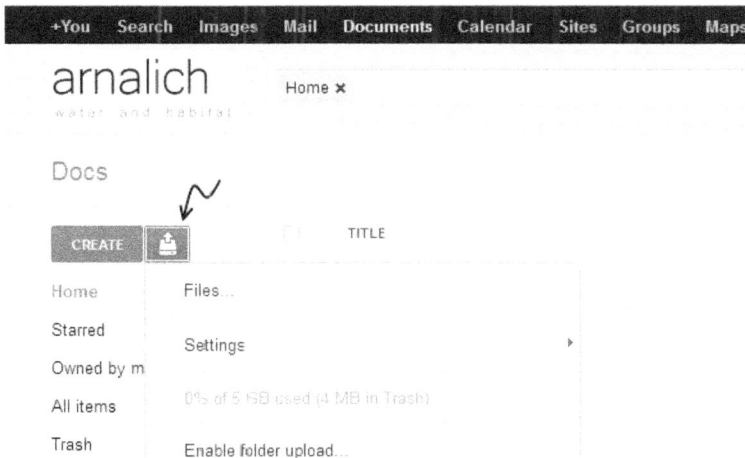

4. In the dialogue that opens it is important to select Convert documents, presentations... If you don't select that option you can upload the file but you won't be able to edit it:

Upload settings

Set your preferences for uploading files. We'll apply these settings to any files

☑ Convert documents, presentations, spreadsheets, and drawings to the co

☐ Convert text from PDF and image files to Google documents

☑ Confirm settings before each upload

Start upload   Cancel

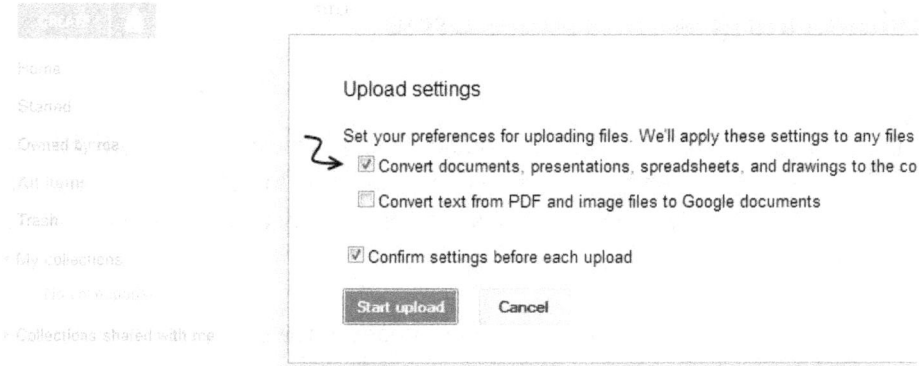

5. Once the file is loaded click on it to open:

Docs

CREATE

Home
Starred
Owned by me
. . .

TITLE

☐ ☆ ▦ meulaboh11

The page will load and it's ready to be edited:

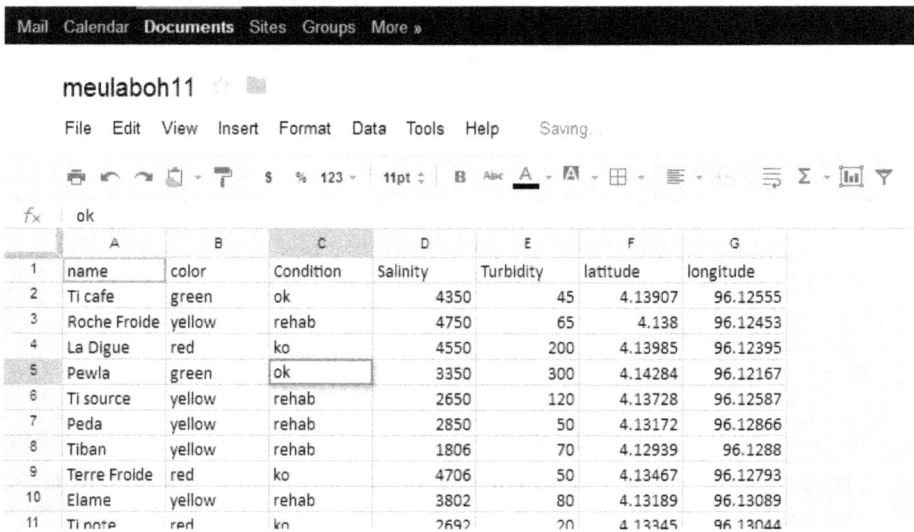

Mail Calendar **Documents** Sites Groups More »

meulaboh11

File Edit View Insert Format Data Tools Help   Saving...

fx | ok

	A	B	C	D	E	F	G
1	name	color	Condition	Salinity	Turbidity	latitude	longitude
2	Ti cafe	green	ok	4350	45	4.13907	96.12555
3	Roche Froide	yellow	rehab	4750	65	4.138	96.12453
4	La Digue	red	ko	4550	200	4.13985	96.12395
5	Pewla	green	ok	3350	300	4.14284	96.12167
6	Ti source	yellow	rehab	2650	120	4.13728	96.12587
7	Peda	yellow	rehab	2850	50	4.13172	96.12866
8	Tiban	yellow	rehab	1806	70	4.12939	96.1288
9	Terre Froide	red	ko	4706	50	4.13467	96.12793
10	Elame	yellow	rehab	3802	80	4.13189	96.13089
11	Ti note	red	ko	2692	20	4.13345	96.13044

Sharing a document

6. To share a document click on share in the upper right corner:

7. Maintain the document as Private and introduce the addresses of the people you want to be able to edit it in the lower box:

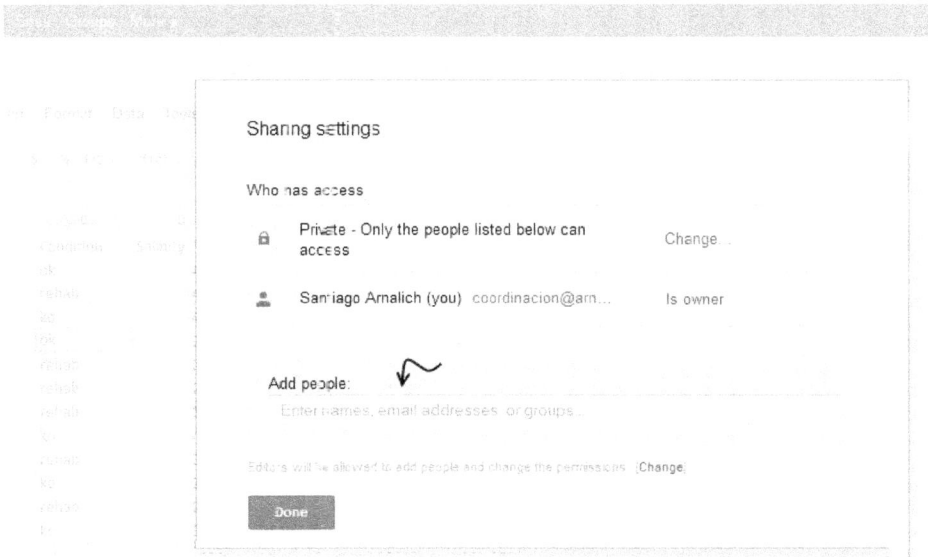

Export as csv

8. Go to > File / Download as / CSV (current page) to download the file in CSV format to use with GPSvisualizer:

That is all, you now have the most important tools to create, share and collaborate using your own maps. We hope that it will be of much use to you and that it brings great benefits to the organizations and the people for whom you work.

Santiago Arnalich

His first contact with GPS was in 1998, during a cycling trip in the Icelandic desert. From thereafter he continued carrying one tirelessly for travel and Development Aid work, exploring all the available technology developments to make creating maps accessible to the average user. That has become a reality with the popularization of Google Earth.

Currently he is the coordinator of Arnalich, Water and habitat. www.arnalich.com.

Julio Urruela

From 2005 he has worked in international aid as a researcher, project coordinator and director of information management. He was initiated to the use of Google Earth in 2006 and in 2010 began training NGO participants of the WASH Cluster in Haiti.

He currently works for UNICEF as a monitoring and evaluation specialist.

Bibliography

1. Arnalich, S. (2007). *gvSIG y Cooperación. Cómo construir e incorporar un Sistema de Información Geográfica a tu proyecto.* Arnalich, Water and Habitat.

    www.arnalich.com/libros.html

2. Crowder, D. (2007). *Google Earth for Dummies.* Wiley.

3. Garmin etrex User's Manual.

4. Google (2012). *Google Earth User's Manual.*

5. Google Earth Blog: www.gearthblog.com.

6. KML support group: http://groups.google.com/group/kml-support/about.

7. MapAction (2008). *Google Earth and its potential in the humanitarian sector: A briefing paper.*

8. MapAction (2009). *Field Guide to Humanitarian Mapping.*

9. MerciCorps (1999). *A Rough Guide to Google Earth.*

10. Puch, C. (2008). *Manual Completo de GPS.* Desnivel.

11. Wernecke, J. (2008). *The KML Handbook: Geographic Visualization for the Web.* Addison-Wesley Professional.

Version 1.0

www.ingramcontent.com/pod-product-compliance
Lightning Source LLC
Chambersburg PA
CBHW051223200326
41519CB00025B/7221